God's Nature and Relation to Creatures

A Catholic Philosopher Explains Divine Attributes, Problem of Evil, and Natural Law Ethics

Dennis Bonnette, Ph.D.

Foreword by Robert L. Fastiggi

En Route Books and Media, LLC
Saint Louis, MO

⊕ENROUTE
Make the time

En Route Books and Media, LLC
5705 Rhodes Avenue
St. Louis, MO 63109

Cover credit: Sebastian Mahfood

ISBN-13: 979-8-88870-239-0
Library of Congress Control Number: 2024946967

Copyright © 2022, 2024 Dennis Bonnette, Ph.D.
All rights reserved.

No part of this book may be reproduced, stored in a retrieval system, or transmitted in any form, or by any means, electronic, mechanical, photocopying, or otherwise, without the prior written permission of the author.

Dedicated to my wife Lois,
our seven children and their spouses,
and our twenty-five grandchildren.

Acknowledgments

First, I thank Dr. Sebastian Mahfood, OP, President of En Route Books, for suggesting that my numerous online essays should be assembled into one volume, based on their related themes. I thank him for thereby preserving in a single book (and now in a series of shorter books developed from it) several years of my online essays, which otherwise might have been easily overlooked in the large archives of websites, or, even worse, permanently lost when those websites were taken down at some future date.

I thank Christendom College for giving me permission to republish all the articles, which originally appeared in *Faith & Reason*. These essays have been given minor revisions as they were updated and incorporated into newer articles in several online journals. I also thank those same online journals, namely, the *Homiletic & Pastoral Review*, *The Postil*, and *Strange Notions* for initially publishing the forty-one essays which appear in this series of books and for their approval of them being republished in book form.

I am indebted to my late friend and colleague, Dr. Raphael T. Waters, for his thought-provoking intellectual dialogue with me for many years and for introducing me to the work of the eminent Australian philosopher and theologian, Austin M. Woodbury, S.M. Fr. Woodbury's extensive philosophical contributions include materials central to philosophical psychology that I employed in my own research on ape-language studies.

I am also indebted to Dr. Waters for founding the Aquinas School of Philosophy after his retirement from Niagara University in 2004, where I resumed my own post-retirement academic teaching in 2010. This school's mature adult students provided an atmosphere that encouraged my continuous writing of peer-reviewed articles as well as such online essays as those published in this present volume.

I thank Dr. Robert L. Fastiggi and the late theologian Dr. James Likoudis for kindly reviewing certain of my online articles before publication and for offering constructive suggestions. I thank Daniel J. Castellano for contributing important philosophical clarifications regarding modern physics. I thank molecular biologist Dr. Ann Gauger for genetic information relating to the possibility of a literal Adam and Eve. I am indebted to the late philosopher Dr. Peter A. Pagan-Aguiar for insights regarding the nature of human freedom.

I also thank the quite competent skeptics, agnostics, and atheists whose many comments respectfully challenged the philosophical and theological propositions I presented on the *Strange Notions* website. In large measure, it is their up-to-date and in-depth intellectual objections which forced me to develop sharper and more decisive intellectual defenses of these truths–arguments and proofs at a level above and beyond that which is normally required in undergraduate or, sometimes, even graduate university teaching.

Lastly, I thank my wife, Lois, for forgiving my computer-wedded absence during the extensive time required for all this writing–after she had hoped I was finally retired from academia.

Table of Contents

Acknowledgments i
Foreword vii
Introduction 1

Chapter 1: How God's Nature Is Known: The Three-Fold Way 9
    #1 - The Way of Causality 11
    #2 - The Way of Remotion 12
    #3 - The Way of Eminence 15
    Conclusion 17

Chapter 2: How Proofs for God Lead to Divine Simplicity 19
    Efficient Causality in St. Thomas' First Way 20
    Efficient Causality in the Second and Third Ways 22
    Proof of Divine Simplicity 25
    Meaning of Divine Simplicity 26

Chapter 3: God: Eternity, Free Will, and the World 29
    God's Immutability and Eternity 29
    Objections to Free Will in God 31
    God Possesses Free Will 32
    Objections Answered 34
    How God's Eternity Relates to the Temporal World 36
    God Remains Immutable as Temporal Events Unfold 38

Chapter 4: How God Can Know and Cause a Universe of Things 41
    Nature of the Problem 41

A Third Way of Existing.. 43
Metaphysical Materialism is Simply Untrue... 46
How Immateriality Enables God to Know Multiple Objects................ 49

Chapter 5: How to Approach the Problem of Evil 51
Why Does God Permit, Or Even Cause, Evil? 53
The Problem of Pain 57
Evil as Part of God's Plan for Man 58

Chapter 6: Hell and God's Goodness 61
The Scandalous Problem 61
One Aspect of the Solution 62
But Why the Pains of Hell? 63
The Specific Solution 64
Man is a Rational Animal.. 65
God's Love 67
But Most People Don't Even Believe in Hell!....................................... 67
The Doctrine of Hell and Free Will 70
How Many Are Lost? 72
Conclusion.. 73

Chapter 7: Theism vs. Skepticism: The COVID-19 Pandemic............ 75
Classical Theism's Defense of God's Goodness 76
The Nature and Role of Chance Events.. 79
Still, Why Does God Enable Free Agents to Choose Evil? 80
Why COVID-19 Does Not Tell Us Whether God Exists...................... 84

Chapter 8: Why Natural Law Ethics is Rational 87
Eternal and Natural Law 88
Man's Last End: Union with God 90
The Basis for Moral Obligation 91

Chapter 9: Abortion Ethics: Natural Law vs. Naturalism 95
Pertinent Thomistic Doctrines 95
Natural Law and Abortion 98
Naturalism's Abortion Stance 99
Hylemorphism vs. Atomism 101
Ethical and Legal Principles 105
Legality 107

Conclusion 109

Foreword

By Robert Fastiggi, Ph.D. Professor of Dogmatic
Theology, Sacred Heart Major Seminary, Detroit, MI

In the conclusion to the present volume, Dr. Dennis Bonnette offers this perceptive description of a dominant intellectual mindset of our day:

> Skepticism today ensnares the minds of many – especially those who have been educated to believe that ancient philosophical and religious belief systems are outmoded and false and that natural science alone offers all possible tentative truths …. Today's dominant intellectual mindset of positivism and materialism begets a worldview that automatically excludes the presence of any truly spiritual realities, and especially, the need for the transcendent God of classical theism.

Of course, there are still many faithful believers in God and the supernatural. In the halls of academia, however, a worldview is often promoted that maintains science and reason exclude traditional religious beliefs. Those who question God's existence are considered the enlightened ones. Skepticism is presented as true wisdom, and religious belief is considered irrational and sentimental.

Skepticism might be growing in the world today, but it is nothing new. There were movements of skepticism in ancient Greece and India. Originally, the word *skeptikos* referred to those who were inquirers. Soon, though, the skeptics became known as those who raised doubts about any claims of knowledge. Some skeptics even questioned whether know-ledge is possible at all. Richard Popkin identifies two forms of ancient Greek skepticism. The first was the academic skepticism of Arcesilas (c. 315–241 B.C.) and Carneades (c. 213–129 B.C.). This form of skepticism held that no knowledge is possible. The second form of skepticism was that of Pyrrho of Ellis (c. 360–275 B.C.). The Pyrrhonians were more agnostic than the academic skeptics. They preferred to suspend judgments on matters that were not certain. By suspending judgment on questions that were not certain, the Pyrrhonians claimed to reach "a state of *ataraxia*, quietude, or unperturbedness." They considered skepticism to be "a cure for the disease called Dogmaticism or rashness."

With the rise of Christianity, both faith and reason were considered to be in harmony rather than in opposition. The great universities of the Middle Ages featured thinkers such as St. Thomas Aquinas (c. 1225–1274), who used rational arguments to demonstrate the existence of God and the credibility of Christian dogmas. The Catholic Church was considered the authoritative interpreter of matters of faith and reason. Confidence in the Church's teaching authority, however, was challenged by the various Protestant movements of the 16th century. This breakdown in Church authority coincided with the revival of skepticism in the late 16th century. This was the period when the *Outlines of Pyrrhonism* by Sextus Empiricus (c. 150–225 A.D.) were translated into Latin and published in 1562. Soon thinkers were raising doubts about whether anything could be known with certitude. Descartes (1596–1650) tried to overcome hyperbolic doubt with his famous "*Cogito ergo sum*" (I think, therefore, I am). The rationalism of Descartes, however, led to other problems.

Philosophical skepticism eventually led to skepticism about religion in general. Anticlerical movements such as Freemasonry led to attacks against the Church during the French Revolution. Atheistic Marxism later resulted in systematic persecutions of religious believers. In democratic regimes of today there might not be official opposition to God, but there is much intellectual and pseudo-intellectual opposition.

It is within this contemporary context of skepticism, materialism, and positivism that the voice of a Catholic philosopher like Dennis Bonnette is so very much needed. Well-trained in Thomistic epistemology and metaphysics, Bonnette is able to apply calm and rational responses to those who raise questions about the existence of God, the freedom of the will, the essential difference between humans and animals, and the possibility of miracles. The chapters in this book for the most part were published as articles dealing with these contemporary challenges to traditional religious belief. Professor Bonnette has provided a remarkable service to those who feel intimidated by contemporary skeptics.

The First Vatican Council (1870) taught that "the one true God, our creator and Lord" can be known with certitude "with the natural light of human reason." This confidence in the ability of human reason to know the reality of God is what animates the chapters of this book. Dennis Bonnette speaks with reason and confidence throughout its pages. Those who feel intimidated by the arguments of skeptics, agnostics, and atheists will do well to read this book. It will provide them with an appreciation of how genuine Catholic philosophy can easily respond to those who criticize traditional theism and the world of the spirit. The Catholic Church upholds the harmony of faith and reason. Dr. Dennis Bonnette is a living witness to this harmony, and he shows how sound reason provides outstanding support to the truths of the faith.

Introduction

Following Vatican Council II, most of the previously orthodox Catholic colleges and universities slowly abandoned their firm commitment to the Catholic intellectual heritage, especially by decreasing both the number and traditional content of required theology and philosophy courses – and even by changing theology courses into what they called "religious studies." This tendency was especially evident in their failure to continue to teach the Thomistic philosophical sciences, such as Aristotelian logic, philosophical psychology, metaphysics, natural theology, and natural law ethics. Such courses were routinely replaced by far fewer ones, which were then taught using an historical method inherently inimical to the truth status of competing historical positions. This, in turn, has led to generations of otherwise educated Catholic college graduates who have little or no real understanding of the Church's intellectual heritage, and especially, its unequalled contributions offered by the philosophy of St. Thomas Aquinas.

In 2022, my book, *Rational Responses to Skepticism*,[1] a rather large volume (569 pages), was published – offering robust rational defenses of the philosophical foundations of authentic divine revelation, specifically as seen in Catholic doctrine. It also contains defenses of certain central revealed truths, such as the genuine scientific possibility of a literal Adam and Eve.

---

[1] https://www.amazon.com/Rational-Responses-Skepticism-Philosopher-Intellectual/dp/B0BQXY8BMH

This present book, *God's Nature and Relation to Creatures,* is the fifth of a series of books containing shorter portions of the much larger volume – each with themes coherent in themselves as subjects of independent investigation, and yet, still subordinate to the overall message of *Rational Responses to Skepticism.*

God's Nature and Relation to Creatures contains nine chapters. Each was itself originally published online as an independent article – making it easier for the reader to make complete sense of its content without reading the other eight chapters or the entire large volume from which it came.

Atheists and skeptics tend to revel in efforts to show that the very notion of God makes no sense at all – that it is entirely incoherent and even self-contradictory. While not entirely omitting discussions of the various classical proofs for God's existence, this book aims primarily at helping readers understand that the nature and attributes of the God of classical theism is both intelligible and coherent. By that, I do not mean to imply that the essential nature of God can be actually understood by living mortals, but merely that we can show that nothing in God's nature is either absurd or entails actual conflict with some other divine attribute.

Chapter one of *God's Nature and Relation to Creatures* first looks at how we come to know the divine nature and attributes by way of the same use of reason employed to discover his existence in the classical proofs for God. This chapter explains the famous "threefold way" by which reason discovers the nature of God: (1) the way of causality, (2) the way of remotion, and (3) the way of eminence.

The "way of causality" argues that, since nothing can give what it does not have, whatever qualities or attributes are found in creatures must also be found in God, since God is the cause of all the qualities found in his creatures. Since non-being does not need a cause, this "way" applies only to *positive* qualities, such as goodness, perfection, wisdom, and being itself.

Still, the "way of remotion" immediately cautions that we must not predicate negative qualities of God, since non-being does not need a direct cause. For example, while the perfection of goodness is directly caused by God and must be found in his nature, evil, which is the *lack* of a due perfection, must not be said of God. That is because all lack of perfection or limitation, which is inherent in creatures, is on the side of non-being, a lack of something that could or should be in a creature – and non-being needs no cause. The way of remotion "removes" all such imperfections from God's nature.

Finally, the "way of eminence" affirms that all good and positive qualities that are properly predicated of God must be infinite perfections, since God is identical with his own essence. This means that any perfection found in God is identical with the infinity of his very being.

These three ways of rationally approaching a proper understanding of the divine nature are all carefully explained and correlated in a coherent manner in chapter one of *God's Nature and Relation to Creatures*.

Chapter two explains and demonstrates what is known as the divine simplicity. Not only is there solely one true God in reality, but that God is intrinsically one and whole – having no composition of parts, principles, or distinct beings. Given all the perfections and qualities attributed to the God of classical theism, one might wonder how they are all found in the same divine being without it entailing some kind of composition or division within God himself.

Chapter two shows that God's essence and existence are one and the same and that all the perfections attributed to him are ontologically one and perfectly identical to the divine essence. God is not composed of potency and act, form and matter, or essence and existence. Simply put, there is no composition in God at all, since, if there were, some prior composing cause would have to effect such composition, and therefore, God would not be the First Uncaused Cause.

The Christian dogma of the Trinity allows a real distinction of relation between the three divine persons, but this can be shown not to entail a real distinction between principles, parts, or things.

Chapter three deals with the immutability and eternity of God and how those divine attributes comport with God having free will. Making this topic even more intriguing is the fact that the eternal and unchangeable God of classical theism creates a world that is ever-changing as it progresses through time.

God's immutability flows from the fact that, as the Uncaused First Cause, God cannot be the subject of motion or change, since what changes or is in motion requires an ontologically prior cause to explain its alteration. But, the First Cause has no prior cause, and hence, cannot be subject to change. This means God is also completely outside time, since time entails progression and change. God must be eternal, that is, completely timeless.

Like all intellectual beings, God has free will. But his free will is identical with his essence, which is necessarily unchanging. So, how can an unchanging will also be free? This chapter explains how God is an eternal pure act of free will, which, while being eternally unchanging, nonetheless eternally wills and knows the ever-changing nature of physical creation.

Chapter four explains how God can know and cause a universe of things. Since some of his creatures possess intellectual knowledge, God must also possess intellect in order to create such intellectual creatures. Because of the divine simplicity, God's intellect is identical with his very essence. But, God's essence is to be the unchanging eternal cause of all finite things. Therefore, his intellect, in the act of knowing all creation, is identical with his essence in the act of causing all finite things. Therefore, God both knows and causes every least aspect of his created world.

Chapter five deals with the problem of evil in relation to the goodness of God. Essentially, if God is all-good, how can he be the cause of a finite world filled with evil? This chapter offers rational solutions to this famous objection to the existence and goodness of the God of classical theism.

The general solution to this question consists in pointing out that God never directly wills evil. God permits evil to exist solely to achieve some greater good end, but never wills the evil for its own sake. This is a complex topic, which is why chapter five considers many diverse aspects of this question.

In fact, chapter six examines what is, perhaps, the most difficult objection to God's goodness: the existence of hell. Hell is understood as a place in which, not only is the sinner eternally denied the happiness of heaven and union with God in himself, but also the sinner is believed to suffer intense and unending physical pain.

Initial considerations entail such notions as realizing that God, as the supreme lawgiver is not bound by the natural laws that bind men. Moreover, it must be realized that the sinner ultimately condemns himself through deliberate misuse of his own free will. Since we not only have intellects, but also senses, the sanction of physical punishments may serve to impel more souls to seek God as their last end than merely the thought of "missing" God. The complexity of this topic is such that one must read the entire chapter in order to realize that its total solution is both coherent and true.

Chapter seven addresses the same problem of evil, but specifically in terms of why God would allow humanity itself to be subject to some great evil, such as a deadly pandemic or world wars.

Chapter eight introduces the notion of natural law that is promulgated by God as the supreme lawgiver.

For those who recognize the existence of God as well as man's possession of a conscience and the ability freely to choose or refuse to live in accord with their human nature as given to them by God, the elements of the natural moral law and its binding force on human acts is evident.

God promulgates his natural laws through (1) the physical laws that govern the activities of all material things and (2) the moral law which binds the actions of rational creatures, such as man. God expresses his intentions by making natures ordered to certain specific ends, such as that a rock naturally tends to fall in a gravitational field or that human sexual acts are ordered to procreation as their last end so as to assure the continued existence of the human race.

Unlike purely physical laws, natural moral laws bind human consciences insofar as we understand the nature of our various actions. For example, since the power of speech serves the end of human communication, deliberate lies are properly understood as violations of the natural moral law.

Finally, chapter nine applies the natural moral law to the question of abortion. Philosophical analysis applied to biological science makes clear that human life begins at conception. Since the natural moral law forbids the direct taking of innocent human life, abortion is morally wrong. This analysis is grounded in a proper understanding of hylemorphic (matter-form) cosmology as first proposed by Aristotle. Naturalism or physicalism fails to grasp the force of the natural moral law, since its atomistic understanding of the material world fails to see the existential unity of things above the atomic level – a critical realization that is key to seeing the unified new life present from the moment of conception, a life whose inherent dignity forbids its deliberate destruction.

These nine chapters of *God's Nature and Relation to Creatures* explore many intriguing aspects of the nature of the God of classical theism – explaining and defending the internal coherence of the divine attributes. They also show how God is related to the created universe, how the problem of evil can be solved, and how God governs all finite beings through natural physical and moral laws – even down to the present need to defend the life of unborn human beings.

Chapter 1

How God's Nature Is Known: The Three-Fold Way[1]

Acceptance of God's existence is conditioned for many on whether a convincing proof thereof can be presented to them. But for others, it is not a problem of proving that God exists, but rather questions about whether the concept of a Supreme Being is even coherent. Many atheists or agnostics simply find the classical conception of God to be unintelligible. God is said to be omnipotent, omniscient, eternal, all good, omnipresent, and so forth. But to many it is not at all clear how these divine attributes can either be proven as real or, more importantly, how they make any sense or can co-exist in one and the same entity or in relation to the world around us. Skepticism of the classical notion of God is evident in the tendency among atheists to deny the existence of *"any* gods," rather than of just the *one* "God."

Moreover, even if one accepts that the classical proofs for God do demonstrate an Unmoved First Mover, an Uncaused First Cause, a Necessary Being, and so forth, how can we prove that these are even the *same* being—or that this being possesses the properties associated with the classical conception of God assumed by most Western philosophers?

---

[1] This article first appeared online on the *Strange Notions* website. https://strangenotions.com/how-gods-nature-is-known-the-three-fold-way/

This article will not attempt to prove God's existence. His reality is merely assumed for present purposes. I have previously offered arguments for God's existence on *Strange Notions* here[2] and here.[3] Still, St. Thomas Aquinas presents the best proofs for God, as found in his opuscula, *De Ente et Essentia (On Being and Essence)*, chapter four, his *Summa Contra Gentiles* I, chapters 13 and 15, and his famous *quinque viae* (Five Ways) of his *Summa Theologiae* I, q. 2, a. 3, c.—these, taken together with their interpretations by classical and modern commentators. Also, I do not intend herein to show how the divine attributes are coherent, consistent with each other, and consistent with the created world in which we find ourselves.

The present enquiry's sole purpose is to show *how* the human mind can come to know the *nature* of God, once his *existence* is demonstrated. If any particular divine attribute is mentioned, it will be primarily to illustrate the methods being explained and not to attempt a full explanation or defense of that attribute's existence or coherence.

Classical metaphysics attains knowledge of God's nature by means of an interpretation, mostly taken from the Christian Neo-Platonist Pseudo-Dionysius the Areopagite (c. late fifth century) known as the *via triplex* or three-fold way.[4] This entails (1) the way of causality (*via causalitatis*), (2) the way of remotion or nega-

---

[2] https://strangenotions.com/how-new-existence-implies-god/

[3] Available online at https://strangenotions.com/how-cosmic-existence-reveals-god%E2%80%99s-reality/

[4] Available online at https://plato.stanford.edu/entries/pseudo-dionysius-areopagite/#OnDivNam

tion (*via remotionis*), and (3) the way of eminence (*via eminentiae*). It is only by understanding how these three methods work and how they interface with each other that it is possible to begin to establish a realistic and coherent understanding of what can be correctly said about the divine nature.

#1 - The Way of Causality

According to the Platonic doctrine of participation, creatures participate in the "pure forms" of heaven by way of imitation. Thus, an earthly horse reminds us of "horseness-in-itself" eternally existing in a spiritual pure form. Christian thinkers transformed this "participation" from mere imitation into real causality, where participation's etymology (cipere—to receive, pars—a part) became ontologically real. Now, in virtue of preexisting formal perfections in the Creator, the creature is directly caused to have its own proper intrinsic form—thus making it to be, say, a horse.

Employing the basic metaphysical principle that non-being cannot beget being, it is self-evident that a being cannot give what it does not have. Certain proofs for God's existence, for example, the second of St. Thomas' five ways, show that he is the Uncaused First Cause. As such, he cannot cause qualities or perfections in creatures which he does not possess. Therefore, any perfection we find in a creature must somehow preexist in God. If a creature lives, then God must be alive. If a creature has intelligence, then God must be intelligent. If there is goodness in creation, then God must be good. If some creatures are persons, then God must be personal. The principle is evident. Still, considered alone it does not give us the full picture of how we form a coherent notion of God.

#2 - The Way of Remotion

What about things we find in creation that do not manifest perfections, but imperfections? Creatures are finite. Does that mean God is finite? There is evil in the world. Does that mean God is evil? Our intellects often make mistakes. Does God make mistakes? There is pain and suffering in the world. Does God suffer pain? The world is constrained by time. Does God exist in time? And in particular, how can a spiritual First Cause create material being? How can that which lacks matter cause matter to exist? These and many other questions arise in which it appears that God is causing problematic effects.

The key to resolving such enigmas is to remove from God any negation, imperfection, limitation, non-being, or evil found in the creature. For any creature to exist, God must create every extent of being or perfection of existence found within it. *But non-being needs no cause.* Hence, causality need not imply that God causes anything that entails limitation or imperfection in creatures.

Perhaps, the most challenging question would be, "How can God cause material things when he lacks it himself? If he cannot give what he lacks, how can God, who lacks materiality, give materiality to physical beings?"

And yet, being material simply isn't all that great! It entails being extended in time and space, which also means to be *limited* by time and space. Being limited in time means *not* to possess the quality of being present to all time at once. Being limited in space means *not* to be present in all places at once. Still, such lack of being in these various respects self-evidently requires no direct causation from God. Moreover, Being material means to be composed of form and matter, which necessarily entails the possibility of decomposition and, thereby, destruction. The alternative would be to create no physical beings at all.

But cannot material beings impact other material beings? Yes, but so can God—merely by creating or uncreating whatever existential qualities are needed to change a thing.

God can create material being because he pre-eminently contains all the existential perfections contained within it, but without the corresponding defects that come with being an actual material substance.

Similarly, the very notion of a finite thing is that it possesses some perfections of existence and lacks others. God is needed to cause what being or perfection is present, but need not be the cause of what is lacking to a finite being. "Finitude" is not a name for the perfection of a being, but a reference of its very lack thereof.

Likewise, evil is not absolute non-being, since that would not exist at all. The proper metaphysical definition of evil is the lack of a *due* perfection, that is, the absence of some property or quality that *ought* to be in a given nature. For example, having a "cold" is the lack of the good health we should enjoy. Yet, the "cold" itself is caused by the multiplication of a virus in us. From the standpoint of the virus, having a "good" cold means the virus is thriving, while we are not!

Moral evil is the performance of an act that deviates from what a human being ought to do so as to attain his last end in God. It is performed to attain a good of some sort, but by a means that is lacking proper ordination to human nature's true end. Again, God is the cause of all that is good in creatures, but moral evil is the result of a use of free will contrary to the good intended by God for human nature. That is to say, the lack of proper ordination of human free acts is caused by man's misuse of his free will, not by an action of God himself. Thus, the moral evil is man's responsibility, not God's. Without addressing all the complexities of the problem of evil itself, the general principle is to remove from predication of God's causality anything in a being that constitutes a mere lack of what is proper to its nature. The nature itself needs God's creative causality; the lack does not.

Always focusing on the being of things makes clear what actually needs a cause and what is merely the lack of what ought to be, as measured by the nature of the thing. The challenge in each case is to determine precisely what positively requires a cause of existence versus what is merely a negation or lack of being. While God is needed to cause the perfection of a nature, its defect or lack need not be attributed to God as the ultimate cause of the thing itself.

What has no need to be caused, namely a lack of being, need not be said of God as its cause—since there is nothing to be explained by a cause. Thus, being is predicated of God; finitude is not. Goodness is predicated of God; evil is not. Intelligence is predicated of God; mistakes are not. Suffering is found in creatures, but not in God. Creatures are constrained by time and space; God is

Chapter 1: How God's Nature is Known: The Three-Fold Way

not. Some creatures are material; God is strictly immaterial, which is the proper meaning of "spiritual."

One can see the nature of remotion or negation in the very way we speak of God as opposed to creatures. We say God is infinite, immutable, uncaused, non-contingent, immaterial, and so forth. In each case, we find a quality of creatures that mark their "creatureliness" and negate its application to God. Each term has a prefix indicating negation, followed by a term marking the finitude of the creature. Thus, we render a judgment that affirms that God possesses some perfection, but in a manner absent the limitation of that same perfection as it is found in the creature.

For example, we do not directly know what the "infinite" in itself may mean, but we do know that God's way of existing is not limited the way that creatures are limited beings. Because of the negative form of the prefix involved, the etymology of some of these attributes may make them sound as if they were something negative. Still, we should remember that such attributes express in fact a positive content.

Some have been deceived by such "indirect naming" into thinking that it is impossible to form any concept of God, saying that he is so ineffable that nothing can be known of him at all. Nothing could be further from the truth, since a term, such as "infinite," actually *affirms* the infinite perfection of God's being.

#3 - The Way of Eminence

Finally, we consider the way of eminence, which is manifest from the conclusions of St. Thomas' proofs for God given in the fourth of his Five Ways and in his *De Ente et Essentia* argument. These proofs conclude to God as greatest in being, a being which is its very act of existing. That is, God is found to be that Supreme Being whose essence and act of existence are absolutely identical.

These arguments show that all the lesser perfections of existence that are found in creatures must be found in God in a manner that is identical with his very essence. Thus, whatever perfection is found in creatures is said to be preeminently contained in God. And, since God's being is infinite, this means that any perfection found in creatures must be found in God as infinitely expressed. We sometimes illustrate this by saying that while man *has* intelligence or goodness, God *is* intelligence or goodness itself.

This again follows from the idea that creatures participate in the divine perfections—receiving, as it were, a part of the divine reality itself.

Here we see a certain conflating of the *via triplex* with the doctrine of analogy. Metaphysical analogy is based on a relation between creatures and God which expresses a real similarity or proportion, but not the exact same meaning of terms used to describe the subject. It is unlike univocal predication. When we say a tiger is an animal and a dog is an animal, the meaning of "animal" is exactly the same. This is univocal predication. But when we say man is a being and God is a being, the term, "being," does not express the same formal identity. The predication is analogous. For in creatures, existence, or being, is received into a nature from which it is distinct. In the proofs of God's existence, creatures are revealed to be effects of God's creative act, which means that they receive their act of existence (*esse*) from God.

Chapter 1: How God's Nature is Known: The Three-Fold Way

In fact, that need to receive existence from an external cause is why the creature needs to be created. But, in God, his very nature is to exist. His existence is identical with his essence. Indeed, every perfection of existence which is found in creatures must exist in God in a manner identical with his infinite essence or nature. Thus, each perfection found in creatures in a limited manner is found in God infinitely expressed, which is precisely the meaning of the way of eminence.

Conclusion

I have not directly intended to explain or defend the divine names or attributes in this article. Yet, by studying the conclusions of the various arguments to God's existence, one can come to see the necessary properties of the divine essence. Thus, through understanding the logical implications of God being the First Mover, First Cause, Necessary Being, Supreme Being, and Ultimate End, knowledge of the various divine attributes, their internal and relational coherence, as well as the intelligibility of God's relation to the world becomes possible. All the while, it remains true that in God these various divine names refer to one and the same identical reality.

While many discussions and disputes arise concerning the coherence of God's nature, the primary purpose of this article has *not* been to defend that coherence, but rather to show the proper method for investigating the divine attributes—the project which logically follows after having discovered that God actually exists.

Therefore, the above explanation is not the end of man's exploration of God's nature, but simply the key to its proper beginning and methodology.

Chapter 2

How Proofs for God Lead to Divine Simplicity[1]

According to the First Vatican Council, the existence of God can be known with certainty by the natural light of human reason through those things that have been created. (*De Fide*).[2] Pope Pius X specified this statement more exactly by affirming that God's existence can be known "as a cause is known with certainty through its effects, from those things that have been made, that is, by the visible works of creation…." (*Sententia fidei proxima*).[3]

Since every being must have a sufficient reason for its being or coming-to-be, an effect is properly defined as any being whose sufficient reason is not totally within itself. To the extent that a being fails to fully explain itself, some other being must be posited which supplies that reason which remains unexplained by the effect. That *extrinsic* sufficient reason is called a "cause." Thus, while every cause is a sufficient reason, not every sufficient reason is a cause. God is his own sufficient reason, but it would be absurd to say that he is his own cause.

Since all human knowledge begins in sensation, it is reasonable that all proofs for God's existence must begin with data taken from sensible creation. This starting point is then shown to be an effect

---

[1] This article first appeared online on the *Strange Notions* website. https://strangenotions.com/divine-simplicity/
[2] Denzinger 1806.
[3] Denzinger 2145.

of a cause—with a possible chain of intermediary causes leading back to an uncaused first cause, which can subsequently be demonstrated to be God.

Efficient Causality in St. Thomas' First Way

While St. Thomas Aquinas' famous five ways to prove God's existence, as presented in his *Summa Theologiae*,[4] employ more than just the efficient, or making, cause (for example, the fifth way is clearly focused on the final cause), demonstration of God's absolute simplicity can be accomplished by focusing exclusively on efficient causality.

The first way begins with the observation that "it is certain and most evident to the senses that some things are in motion." As has been proven in an earlier article,[5] "whatever is in motion must be being moved by another." That "other" is a cause of motion or coming-to-be (*causa fieri*), which cause may be either an efficient, or making, cause–or, it may be a final cause. While modern readers of the argument from motion quite naturally tend to think of the movers as efficient causes of motion, it may come as a surprise to some to learn that Aristotle had in mind also final causes–so that his unmoved first mover moves things in motion by means of *attraction*, not efficient causality.

The modern understanding of causality as it takes place in motion tends to be influenced by David Hume—so as to think of it as

---

[4] *Summa Theologiae*, I, q. 2, a. 3, c.

[5] Available online at https://strangenotions.com/whatever-is-moved-is-moved-by-another/

sequences of "events" in which prior ones causally influence subsequent ones. But, such "causality" does not meet metaphysical criteria, since a delay of even a nanosecond between cause and effect would entail that the cause might be non-existent by the time the effect is produced. Clearly, an effect, which is deriving some existential perfection from an efficient cause, cannot be receiving it from a non-existent cause. Hence, the cause as causing and the effect as being effected must be simultaneous. Thus, efficient causes of motion must be simultaneous with the motion they cause in another.

In both the first and second ways, St. Thomas affirms the principle that there can be no infinite regress among intermediate causes, which is evident in that no intermediate cause is a fully sufficient reason for its own effect, which is the reason it is called an "intermediate cause." Were all causes intermediate, then, regardless of number, the complete sufficient reason for the final effect would never be fulfilled—which is impossible. The impossibility of an infinite regression among proper causes has also been demonstrated in an earlier article.[6]

Moreover, motion entails the production of "new existence" with respect to the thing being moved, so that it is not merely "motion" that the unmoved first mover causes, but the very existence of the new perfections of existence manifested by any change in being. The unmoved first mover is an efficient cause of new existence in all things in motion, even if that new existence is merely in the

---

[6] Available online at https://strangenotions.com/why-an-infinite-regress-among-proper-causes-is-metaphysically-impossible/

order of accidental being in the Aristotelian sense. The need for a "universal donor" of new existence has also been demonstrated in an earlier article.[7]

Moreover, potency is what is able to be, but is not; act is what actually exists. Thus, motion is the progressive actualization of potency. Since things in motion must be moved by another, and since no infinite regress of moved movers is possible, there must be a first mover in which no motion occurs. But the total absence of motion means that the unmoved first mover acts to cause motion, and yet has itself no potency being progressively actualized, that is, it is pure act as the efficient cause of motion in things.

This unmoved first mover must also be the "universal donor" of new existence, since both entities have the identically same role in accounting for the coming-to-be of all the new existence manifested through motion in the world.

Efficient Causality in the Second and Third Ways

While the first way deals with causes of coming-to-be (*causa fieri*), the second way deals with causes of being (*causa esse*). Since modern physics tends to challenge the simultaneity of macroscopic examples of such causation, suffice it to point out that (1) unless simultaneity existed in causes of motion, no motion could occur, since a "past" mover cannot "presently" move something, and (2) it is possible that the second way immediately enters the metaphysical

---

[7] Available online at https://strangenotions.com/how-new-existence-implies-god/

order with causes that sustain existence which immediately transcend the physical order. As shown in a previous article,[8] the very existence of the cosmos requires an infinitely powerful Creator.

Whether we consider efficient causes of coming-to-be or of being, it really does not matter, since what is absolutely evident is that, unless causes exert causation through *immediate and direct* influence on the effect, no effect can be produced at all—for the same reason that an effect needs a cause in the first place, namely, an existential need in the effect must be met *here and now* by an actually acting cause.

In a proof for the eternity of God that is found in his *Summa Contra Gentiles*, St. Thomas takes as his starting point things whose existence or non-existence is possible. He argues, "But what can be has a cause because, since it is equally related to two contraries, namely, being and non-being, it must be, if existence accrues to it, that this is from some cause."[9]

The third way of the *Summa Theologiae* is far too complex to treat in detail here, but I have explained it more fully elsewhere.[10] It is not an argument from the contingent to the necessary, as it is so often mischaracterized, but rather an argument from the possible and necessary to being *per se* necessary. Using the notion of the possibles as expressed in the *Contra Gentiles*, it is evident that not all beings can be merely possible beings, since possible beings are

---

[8] Available online at https://strangenotions.com/how-cosmic-existence-reveals-gods-reality/

[9] *Summa Contra Gentiles*, I, ch. 15, para. 5.

[10] *Aquinas' Proofs for God's Existence* (Martinus-Nijhoff: The Hague, 1972), pp. 127-139.

caused beings and no infinite regress among proper causes is possible, as has been shown.[11] There must be an uncaused first cause in any regression of possible beings, and that first cause cannot itself be another possible being, since all possible beings are caused.

Hence, some being must exist whose existence is not merely possible to be or not be, but rather must necessarily exist. St. Thomas then traces from necessary beings that receive their necessity from another to that being which is necessary through itself, namely, God—again, since "it is impossible that one should proceed to infinity in necessary things which have a cause of their necessity, as has been already proved in regard to efficient causes."[12] This necessary being must account, not only for its own necessary existence, but also for the existence of all other things—both necessary and possible, as defined in the third way.

All the above has been intended simply to show that some of the classical proofs for God's existence demonstrate that an uncaused first cause must exist and, as St. Thomas observes, that that first cause fulfills the nominal definition of the classical meaning of God. Moreover, this first cause must cause the very existence of the cosmos—in both the substantial and accidental orders.

Just how well this uncaused first cause fulfills the classical definition of God depends upon our understanding of its nature.

---

[11] Available online at https://strangenotions.com/why-an-infinite-regress-among-proper-causes-is-metaphysically-impossible

[12] *Summa Theologiae*, I, q. 2, a. 3, c.

Proof of Divine Simplicity

The purpose of this paper is to determine whether the first cause meets one of the essential attributes of the classical meaning of God, namely, the divine simplicity. Divine simplicity means that God is not composed of parts, principles, or things.

The critical importance of establishing that God is the absolutely first cause, not only of the coming-to-be of things, but of their very existence, is that being a first cause of existence precludes any form of composition in God.

St. Thomas makes two clear points here. First, God is truly and absolutely simple "because every composite is posterior to its component parts, and is dependent on them; but God is the first being" and second, "because every composite has a cause—for things in themselves different cannot unite unless something causes them to unite. But God is uncaused, … since he is the first efficient cause."[13]

As the first efficient cause, God can have no prior cause to combine any principles, parts, or things in order to make him a composite whole. Since a composite presupposes the prior component parts that make it up, the composite would then depend on those prior parts. God, as absolute first cause, can depend on nothing prior to himself. Hence, he cannot be a composite of any type.

Again, any composite requires some principle of unity. If that principle comes from without, then the composite cannot be the first cause, since something is prior to it. If either component part accounts for its correlative component part, then they cannot be

---

[13] *Summa Theologiae*, I, q. 3, a. 7, c.

distinct parts—since nothing can give what it does not have—in which case there would be no composition, but only identity.

Moreover, what is composite is made up of diverse components, and diverse things can only be united by some causal agency. But, God, as the first cause, has no cause. Therefore, composition in God is impossible.

Meaning of Divine Simplicity

This means that God cannot be a composite of any potential principle and active principle, such as primary matter and substantial form, or substance and accident, or essence and existence.

The deepest truth about divine simplicity is that in God essence and existence are identical. God cannot have an essence to which is added existence, for whatever is found in anything either flows from its essence or comes to it from some extrinsic cause.[14] But God is the first cause, and so, his existence cannot come from some extrinsic cause. The only alternative is that his existence comes from his very essence. But nothing can give what it does not have. Therefore, God's essence must already contain its very existence. His essence is identical with his existence. This is as simple as any being can be, since in all created things, existence is caused—meaning that existence is something added to essence. But in God this is not the case. He simply is his own act of existence.

God is a pure act of existence—infinite in virtue of that act being received and limited by *no* essence—Pure Act limited by noth-

---

[14] *De Potentia*, 10, 4.

ing at all. Pure Existence limited by nothing constitutes the Infinite Being.

This means that God is not composed of form and matter. Hence, he is not a material body. In him, there can be no composition of substance and accident. Whereas in man, will is distinct from substance; in God, they are one. Nor is intellect distinct from substance. But if God's will and intellect are identical to his substance, then his intellect is also identical to his will. So, too, his acts of willing are identical to the divine substance which is identical to his acts of knowing, making his willing and his knowing to be one and the same.

The various distinctions between substance, faculties, and acts found in man arise because of the composite nature of his being. First, as a creature, we are composed of essence and existence. As a material being, we are composed of substantial form and quantified primary matter. Having accidental qualities that change through time, we are composed of substance and accidents. Our powers are distinct from their operations. And potency is distinct from act in each of these composites.

But, in God—as the absolutely first cause of all things, who himself is caused by nothing–who is his own sufficient reason for existing and being what he is in every way possible–in the one true God, *all these creaturely distinctions are obliterated.*

That is the meaning of the divine simplicity: Pure Existence, with no limiting essence and no real distinctions in God between principles, parts, or things.[15]

There are many other divine attributes, such as God's perfection, goodness, infinity, immutability, eternity, omniscience, omnipotence, and so forth, which cannot be addressed in this short article, and would therefore be off topic for consideration here.

---

[15] The Catholic dogma of the Trinity allows a distinction of relation to exist between the three divine Persons, but this does not entail a real distinction between principles, parts, or things. Such knowledge of God pertains to sacred theology, not metaphysics or natural theology *per se*. It is the proper work of the philosopher to show that such relations do not entail any contradictions in being, but that analysis does not belong to the natural knowledge of God, which prescinds from divine revelation.

Chapter 3

God: Eternity, Free Will, and the World[1]

Rather than present a systematic defense of all the divine attributes involved in this article, my purpose here is to explore some philosophical doctrines about God whose interrelationship appears perplexing, if not outright contradictory—drawing on whatever elements of natural theology are needed. Starting with a proof of God's immutability, I will then consider his eternal life and how it is possible for him still to have free will. Finally, I will consider how it is possible for an eternally unchanging God freely to create and interact with a temporal world that constantly undergoes change. Since some claim that this entire metaphysical scenario is radically incoherent, careful philosophical explanation is mandatory.

Some of the logical steps entailed in this topic are fairly straightforward. Understanding the inferred metaphysical concepts is somewhat more challenging.

God's Immutability and Eternity

As has been shown previously,[2] a key inference of St. Thomas Aquinas' proofs for God's existence is that God is the Uncaused First Cause. Since God is uncaused, he cannot be the subject of mo-

---

[1] This article first appeared online on the *Strange Notions* website. https://strangenotions.com/god-eternity-free-will-and-the-world/

[2] Available online at https://strangenotions.com/divine-simplicity/

tion or change, because whatever is moved or changed must be moved or changed by another.[3] Hence, God is immutable.

Moreover, the Uncaused First Cause must be pure act, since change would require moving something from potency to act. But, if no change is possible, God must have no potency to further act. Hence, he is pure act, which means pure being. In fact, as the absolutely simple first being,[4] God is not even composed of essence and existence. He is pure act of existence without any limiting essence, that is, the Infinite Being. Only one such being is possible, since if there were two, one would limit the infinity of the other.

Some, confusing activity with motion, misconstrue God's immutability as meaning frozen, static, lifeless, and impotent. Quite the contrary, the Infinite Being already possesses all existential perfections so completely that change could give no greater activity or power.

God's immutability entails his eternity, since what is immutable has neither beginning nor any progression through time. *God is utterly outside of time*, existing as it were "all at once." Ordinary language betrays human understanding of God's eternity. Eternity does *not* mean endless duration: time without beginning or end. God's eternity means the simultaneous and complete possession of infinite life. It is the term defining the divine life of God. We know God is living since he is the cause of that positive existential perfection that we call "life" in creatures. The term, "life," in God must be

---

[3] Available online at https://strangenotions.com/whatever-is-moved-is-moved-by-another/

[4] Available online at http://strangenotions.com/divine-simplicity/

understood analogously in that he does not live with the limitations inherent in earthly organisms,[5] but rather possesses preeminently whatever positive perfections life entails in created living things.

In the divine eternity, God experiences no succession of events. Because of divine simplicity,[6] God's knowledge of himself and, thereby, of the world he causes, is one with his singular causal act whose *multiple objects* are the unfolding sequence of temporal world events—events novel to us, but not to God. God cannot change his mind or will or any aspect of his being during his eternal existence.

Objections to Free Will in God

For us, free will entails considering various alternatives, knowing we can choose one as opposed to others, and then finally, making a choice one way or another. This process takes place through time.

But, God is not in time. He cannot choose between alternatives as we do. Since to choose freely requires that there be a real difference between the potency to various alternatives and the actuality of choosing a single option, time is needed to make the choice. God's eternal immutability appears to preclude him having free will.

---

[5] Available online at https://strangenotions.com/how-gods-nature-is-known-the-three-fold-way/

[6] Available online at https://strangenotions.com/divine-simplicity/

Again, if God is pure act, there can be no distinction between potency and act, meaning that there is no real distinction between what God can do and what he actually chooses to do. Since a thing's nature determines what it is able to do, it would appear, then, that God's nature must determine both what he is able to do and what he actually chooses, since there is no distinction between them. Hence, God's alleged "choices" appear to be determined by his nature, and thus, not free choices at all.

God Possesses Free Will

Still, since the positive perfection of intellect is found among creatures, God must possess intellect–for God could not create finite intellects unless he possesses that perfection himself. Just as the intellect *knows* being as the true, the intellectual appetite *desires* being as the good. The intellectual appetite is called "will." Thus God must have will as well as intellect. In fact, the divine simplicity requires that his will is identical with his intellect.

It may seem odd, but it is possible to have a will that is moved necessarily toward certain objects. For example, God wills his own goodness necessarily. As St. Thomas Aquinas puts it:

"The divine will has a necessary relation to the divine goodness, since that is its proper object. Therefore, God wills the being of his own goodness necessarily, just as we will our own happiness necessarily...."[7]

---

[7] *Summa Theologiae* I, q. 19, a. 3, c.

Chapter 3: God: Eternity, Free Will, and the World

Thus, the notion of will itself, as the intellectual appetite for the good, is not inconsistent with an *absence of free choice*.

And yet, despite being utterly immutable and eternal, God does possess *free* will with respect to some things. While he necessarily wills those goods that are equivalent to his own being, such as his own existence and his own goodness, he nonetheless *does not necessarily will lesser goods than his own goodness*, such as his will to create this world or that world or not to create at all.

Again, St. Thomas explains:

"God wills things other than himself only insofar as they are ordered to his own goodness as their end. ... Hence, since the goodness of God is perfect and can exist without other things, inasmuch as no perfection can accrue to him from them, it follows that for him to will things other than himself is not absolutely necessary."[8]

St. Thomas maintains a *suppositional necessity* here, saying, "… supposing that God wills a thing, then, he is unable to not will it, since his will cannot change."[9]

The immediate evidence of the existence of such freedom by God to will lesser goods than himself is the evident fact that the finite world in which we live actually exists, as opposed to an unlimited number of possible other worlds he could have created. Is he necessitated to create this world? No, because this world is a

---

[8] Ibid.
[9] Ibid.

lesser good than his own goodness which already includes every possible perfection of goodness. Hence, God creates this finite world in which we live by a perfectly free act of his will.

Objections Answered

First, some think that God being the Necessary Being is inconsistent with the contingency of his free will choosing to create this world, which did not have to exist at all. Although God is the Necessary Being, this necessity refers primarily to his act of existence, since his essence is identical to his existence[10]—thus, making it impossible for him not to exist.

The term, "necessary," with reference to the divine nature *cannot* be capriciously defined to suit some contrived antitheistic argument. Its meaning originates in the context of St. Thomas' Third Way, which refers solely to a being whose necessity for existence comes from itself and not from another.[11] Such a being must be that being whose essence is its very act of existence.

Hence, God's necessity means primarily the necessity of his existence. As shown by St. Thomas above, that necessity also pertains to God's willing his own goodness, since it is equivalent to his own being—*but it is not necessary for God to will things other than himself.*[12]

Thus, when God chooses freely to create this world as opposed to any other, this choice does not make him to somehow become a

---

[10] Available online at https://strangenotions.com/divine-simplicity/
[11] *Summa Theologiae* I, q. 2, a. 3, c.
[12] *Summa Theologiae* I, q. 19, a. 3, c.

"contingent" being. He is still the one and only Necessary Being, but he makes a free choice that in no way contradicts his existential necessity.

Second, some object that God cannot have free will, since that would necessarily entail a change in him, which his immutability and eternity forbid. But this is to make the gross error of thrusting God into time—as though he was first not making a choice and then later making one, which would be a change in him.

Unless one misconceives God in a material, temporal fashion, the metaphysical insight required is to grasp that God's very substance is an eternal act of will in which some objects are willed necessarily and others are willed non-necessarily. This is not an act having temporal duration in which choice begins at some point. God is simply his own act of choosing—a choice eternally identical with his very substance through divine simplicity.[13]

Third, it was objected that God's choices are not really free, because his choice is identical to his nature, and therefore, is determined by his nature. It is true that God's nature determines what he is able to do and that his actual choice is identical to that nature. But, *this will prove to be unproblematic.*

While God might have made other logically possible choices (and there might be other logically possible Gods), *such hypothesized alternatives are not metaphysically possible*—given that the one and only *actual* God, who is immutable, has made the choice he has *actually* made. These hypothesized alternatives may be metaphysically possible in an absolute sense, but they are not so *de fac-*

---

[13] Available online at https://strangenotions.com/divine-simplicity/

to—given that only one God actually exists and has made the actual choice he has eternally made.

What is *de facto* metaphysically impossible renders the alternative "logical possibilities" not logically possible at all, except as contrary-to-reality mental imaginings. That is, *they are not actually real possibilities at all.*

God is actually able to do only what he actually freely wills to do, since on the supposition that he wills a certain choice from all eternity, that will cannot be changed–because of the divine immutability. Thus, there is, in fact, no distinction between what God is able to do and what he does do–but *what he does do, he does freely with respect to goods that are less than his own goodness.*

Given the divine nature, God is determined to will his own existence and goodness necessarily. But, he is also determined to will lesser goods than his own existence non-necessarily, which means that *he is determined by his own nature to act freely.* That is to say, with respect to the willing and creation of lesser goods than his own goodness, God is *determined* to be *not-determined.* His nature determines that the divine will's act with respect to certain specified objects, such as the creation of this particular world, is *not necessary*, and therefore, is perfectly free.

Thus is resolved the problem of God's nature "determining" his choice.

How God's Eternity Relates to the Temporal World

Finally, while God's general relation to the created world is a topic far too vast for this article, the question logically arises as to

Chapter 3: God: Eternity, Free Will, and the World

how an eternal, unchanging God can cause the dynamic, changing world we inhabit—without being subject to change himself.

God is utterly outside the created world—existing in timeless eternity. But, according to Christian revelation, the world had a temporal beginning. Moreover, physical creation is subject to constant change and motion. Indeed, that very coming-to-be is the starting point for the most famous proof for God's existence, St. Thomas' First Way.[14]

Some argue that every change in the temporal world requires a change in God to initiate that new causation that changes the world. For, how can one thing initiate new motion in another without itself changing in the very act of "sending forth" its causal influence to the world?

Such reasoning may make perfect sense to a mentality mired in philosophical materialism. But, it makes no sense at all in existential metaphysics. Physical agents change as they cause effects. But to think that this also applies to spiritual agents is absurd and illogical.

Since whatever is in motion or is changed must be moved or changed by another,[15] maintaining that a cause cannot cause change without itself changing would entail an infinite regress among simultaneous caused causes and make impossible an Uncaused First Cause. This is because it would mean that every cause would be an intermediate cause in need of a prior proper cause. If every cause has a prior cause, any causal regress among proper

---

[14] *Summa Theologiae* I, q. 2, a. 3, c.

[15] Available online at https://strangenotions.com/whatever-is-moved-is-moved-by-another/

causes would have to regress to infinity. But, I have shown elsewhere that an infinite regress among simultaneous proper causes is metaphysically impossible.[16] For one thing, the sufficient reason for the final effect would never be fulfilled. Therefore, it is manifestly false to claim that every cause must itself change in order to cause a change in another.

Causality in metaphysics is simply a subdivision of the principle of sufficient reason. The notion of causality arises from metaphysical analysis of the effect, not of the cause. If every being must have a sufficient reason for its being or coming-to-be, then either a thing is completely its own reason for being, or else, to the extent that it does not completely explain itself, something else must. That "something else," or extrinsic sufficient reason, is what we call a "cause."

Thus, the causality principle states that every effect requires a cause. What is changing or in motion fails to explain its own coming-to-be, and hence, needs a cause. Nothing in this explanation of causality logically implies a change in the cause as causing—*only something happening to the effect.*

God Remains Immutable as Temporal Events Unfold

Furthermore, since change takes place in the effect, not in the cause as such, there is no problem with God being eternally un-

---

[16] Available online at https://strangenotions.com/why-an-infinite-regress-among-proper-causes-is-metaphysically-impossible/

changing, while the world could have a beginning and events unfold sequentially in it throughout time.

God, in a simple eternal act of will, causes all events in physical creation to take place at their appointed times. All beginnings and changes take place in creatures, not God. *Indeed, time and space themselves are part of the world's created limitations.* If Christian belief that the world began in time is true, God simply willed the creation of the world to be with a beginning in time–again, something happening to the world, not to its timeless Creator.

This article is not the whole of natural theology. And yet, it does explain how God can be changeless and eternal, while still having a free will through which he causes the sequential unfolding of events in a temporal world of which he is not part—but reigns as its sole and timeless Creator.

Chapter 4

How God Can Know and Cause a Universe of Things[1]

Nature of the Problem

God is absolutely simple, meaning that he is not composed of parts, principles, or things.[2] He is a spiritual being, since what is physical is subject to motion and God, as Unmoved First Mover, cannot be subject to motion.

It seems unimaginable that a simple, pure spirit could both know and cause the nearly infinite myriad of things that God has created. Yet, it is demonstrable that he causes each creature and knows each one individually.

That God causes all finite things follows from the proofs for his existence, since the arguments run from finite effects back to the infinite cause, which is God. Since (1) every finite being is actually being created as it is sustained in existence, and since (2) infinite power is needed to create anything, the First Cause must have infinite power.[3] Infinite power resides solely in an Infinite Being. Were there two such beings, one would limit the power of the other.

---

[1] This article first appeared online on the *Strange Notions* website. https://strangenotions.com/how-god-can-know-and-cause-a-universe-of-things/

[2] Available online at https://strangenotions.com/divine-simplicity/

[3] Available online at https://strangenotions.com/how-cosmic-existence-reveals-gods-reality/

Since only one being has the infinite power needed in order to create things, it follows that all finite things are created by that single Infinite Being.

God, though perfectly simple, somehow creates untold numbers of finite things. Yet, it seems utterly counterintuitive that an absolutely simple First Cause could produce nearly infinite effects either in a single act or in multiple acts that cause the unimaginable multiplicity of creatures.

God's intellectual nature is manifest from the fact that some creatures have intellectual abilities. Since God cannot cause intellectual perfections that he himself lacks, God must be intellectual.[4] And, if God creates all things, he must know what he creates. Still, if God knew creatures the way we know things, his knowledge would depend on observing them. But, the Uncaused First Cause cannot depend on another for anything. He is his own sufficient reason for existing and acting. Therefore, God cannot know creatures by observing them as we do. Rather, it must be that God knows himself as Creator of all things–thereby knowing every least detail of everything to which his creative causality extends.

Were God simply another material entity, such omniscience would be impossible. God's spiritual nature will be seen as the key to how he can cause and know an infinite myriad of things–in a single act of knowledge that is identical to his act of creating.

---

[4] Available online at https://strangenotions.com/how-gods-nature-is-known-the-three-fold-way/

A Third Way of Existing

There are three ways that things can exist: (1) materially, (2) spiritually, and (3) in an intermediate state between matter and spirit. Following the father of modern Western philosophy, Rene Descartes,[5] most people think in terms of things being either material or spiritual, with no third alternative being possible. By "matter," is meant that which is extended or locatable in space. This would include physical forces and energy fields. By "spirit," we understand that which is neither extended nor locatable in space *and is utterly independent of matter.*

But, what if something that is *not extended in space* is still *dependent on matter* for its existence? Such a thing would constitute the third alternative described above: an intermediate state between physical matter and pure spirit.

Examining the sense of sight shows that something belonging to this third category of reality actually exists. When something is seen, it is seen as a whole: top and bottom, left to right. Thus, when I see a tree, I see the whole tree in a single act of sight. If my perception were not so unified, I could never know a whole tree–only tiny, unconnected, and unintelligible "pieces." But, the tree itself is extended in space, and is seen by me as so extended. Can a purely physical entity "apprehend" the whole of anything in this unified fashion? No, it cannot.

Consider a TV screen's image of that tree. Hundreds of thousands of pixels create the image of the tree. Yet, no single pixel con-

---

[5] Available online at https://plato.stanford.edu/entries/descartes/

tains the whole image. Different pixels illuminate to represent different parts of the whole image. But, the viewer sees the whole image all at once. That is one reason why TV screens don't see their own images. Yet, a dog, bounding into the room, instantly sees the tree on the screen–*as a unified whole. Indeed, it can see many trees at once.*

Every physical device "apprehending" an external sense object entails reception and storage of data on a physically extended medium, such as a CD, DVD, monitor screen, tape, chip, microchip, nanochip, or some such entity. In every one of these devices, data is stored or displayed with one part representing one part of the external object, and a distinct part representing another distinct part of the object.

Nothing represents the whole as a whole–for the simple reason that it is physically impossible.

The only way to get the whole image on a TV screen as a whole would be to collapse the vertical and horizontal dimensions to a single "dot" in the center of the screen—such as old picture tube TV's did when turning them off. Now you have perfect unity— only you have lost your picture, since all the photons are hitting the same spot! Analogously, the same logic applies to every other medium of data reception or storage: reducing the data to perfect unity would entail so overlapping data upon itself as to render it meaningless.

This same analysis applies to all forms of sensory knowledge, whether sight, sound, taste, touch, smell, or any other possible form. Because sensory data, by definition, is extended in space, it is impossible to receive or record physically complex data on a single unitary physical "pixel" (for want of a better term).

Note that, although a material entity cannot know the "whole" of anything, a dog that watches television can see an image of another dog on the screen and bark at it! The reason is simple. While the body of the dog (including its brain) may be physically extended in space, nonetheless, its apprehension of the image on the screen is received as a "whole" solely because the dog has an immaterial soul with immaterial sense faculties, which enable him to see the image as a whole. TV screens and other physical representational devices know nothing at all, since they cannot *"take in"* the *wholeness* of sense objects, *which alone constitutes real knowledge of things.*

What is physically extended in space is inherently multiple, since it has parts outside of parts. This is why a physical entity can never "express" the wholeness of another physical entity in a single "pixel" of its makeup. Rather, part A must represent part A of the object represented, and part B represents part B, and so forth—but no single part represents the whole as such. No single physical part *apprehends the whole* all at once.

This would work similarly for senses other than sight. Various electronic "sensing" and "recording" devices designed to detect external sense objects of various senses will require diverse technological mechanisms. Still, in each case the discrete physical parts of whatever physical medium data is held on will necessarily face the intrinsic limitation that each physical element can only represent a single bit of information (probably in binary form), while no single bit unifies in a single act of "apprehension" the entire sense object it represents. Even a collectivity of bits explains nothing, since each bit represents only a part of the whole, and nothing represents the whole all at once.

Nor can one evade this logic by avoiding crude images of atomic entities in favor of esoteric notions of physical forces or energy fields–since the essence of *any* physical reality entails extension in space, wherein the *same* problem arises of discrete parts representing discrete parts of the object known, but nothing adequately representing the unified whole.

Metaphysical Materialism is Simply Untrue

Only an *immaterial* cognitive faculty, that is, *one not extended in space*, can actually apprehend the wholeness of any sensed object. Moreover, in the same act, the sense faculty can apprehend *many* individual wholes at once, as in a flock of birds.

How *does* an immaterial sense faculty unify the object of perception into a meaningful whole? Knowing how an immaterial entity "works" would require knowing how to make one—something that exceeds human capabilities. Still, I know a sense faculty *can* do it, because I actually sense meaningful wholes in sensory experience. That is, in a single act, I *see* a whole moose or experience *hearing* a complete melody or am aware that I am *touching* the total surface of a sphere.

No purely physical entity can *adequately* explain this fact.

Sight's ability to apprehend its object as a whole is sufficient to show that at least one external sense faculty must be immaterial. Because an animal's sensitive soul is immaterial—that is, because it

Chapter 4: How Can God Know and Cause a Universe of Things 47

is not extended in space, even animals can experience the unified wholeness of sense objects—*and many such wholes simultaneously.*

Purely materialistic metaphysic's essential problem is that sense cognition's *immaterial* nature is what enables the knower to apprehend the physically extended object as a unified whole. In so doing, *immaterial cognition achieves something that mere extended matter cannot do, namely, it can unify in a single simple act what in physical reality is extended in space and multiple in parts.*

Some materialists admit that certain cognitive acts cannot be expressed in purely material terms. Yet, they insist that these "epiphenomena" somehow "emerge from" purely physical matter. That is, they are simply a product of physical matter in some way. The problem with this explanation is that the more perfect cannot be explained by the less perfect. Or, to put it another way, *that which is inherently unable to explain the unity of the whole (discrete physical parts) cannot be a sufficient reason for apprehending the thing sensed as a unified whole.*

Moreover, this immaterial principle must explain how unity is achieved from multiple sense data. Since a material entity can never explain the unity of its discrete elements, what unifies must not only be immaterial, but must be something within the sentient organism that *unifies its discrete material organs into a functional whole respecting sense perception.* Such an immaterial principle would be the *form* or *soul* of even the lowest sentient organisms.

This means that a purely materialistic explanation of all reality is simply false.

Since neither individual material parts nor their collectivity can explain the unity of the whole which is sensed, it is clear that material physical components of organisms cannot explain the unity of sensation experienced. This argues to some principle of unity that enables the entire organism to act in a manner which none of its material parts or their collective whole can explain. An immaterial principle of unity is needed, such as the substantial form, which functions as an organizing principle of matter according to Aristotle's hylemorphic (matter-form) theory.[6]

Now, I am *not* saying that the immateriality required for sensation is the same as the strict immateriality of a spiritual soul. For sense knowledge remains dependent on matter to a certain extent, as evinced by the fact that all such knowledge, even in the imagination, is received "under the conditions of matter." That is, we sense a tree as extended in space, having weight, color, shape, and so forth. This indicates that sensory knowledge is still dependent on material organs of a material substance, even though the actual sensing faculty must belong to a soul that is not extended in space.

Yet, it remains true that these sensing powers cannot be explained merely in terms of lower physical units, as shown above. Rather, this is one of those "third alternative" cases of something that exists in an intermediate metaphysical state between physical matter and pure spirit. Most importantly, what is clear is that these immaterial sense powers (1) cannot be explained by metaphysical materialism and (2) possess the *immaterial* quality of being able to unify in a simple cognitive act that which is extended in space and

---

[6] Available online at https://www.britannica.com/topic/hylomorphism

multiple in parts—and even a multiplicity of wholes simultaneously, as when a dozen eggs are seen at once.

The existence of such "intermediate" forms obviously comports with Aristotelian-Thomistic hylemorphic doctrine, but *not* with the materialist claims of some form of <u>atomism</u>. The key insight is that an immaterial cognitive power can manifest what pure physicalism cannot explain, namely, conscious apprehension, in a unified act of cognition, of multiple objects perceived as wholes–as when we see a stand of trees.

While an analogous and even more striking case can be made for the spirituality of the human intellectual soul, I have studiously avoided this topic for two reasons: (1) it would entail explanation far exceeding this article's space limitations, and (2) it can be easily demonstrated that *immateriality in sense cognition enables even a dumb bunny to do something that metaphysical materialism can never explain*, namely, to know in a single, unified act the whole of a sensed object, such as an entire carrot—or even a bunch of carrots all at once.

Such immateriality is the basis for the ability of a single knower to know multiple objects in a single, simple unified act of knowing.

How Immateriality Enables God to Know Multiple Objects

What has all this to do with God's ability to know and to cause the near infinite multiplicity of the created world? Simply this. While we do not know exactly how the immateriality of God's or man's cognition enables them to know multiple, whole objects, or even how animals do it at their own merely sentient level of cognition, still, the fact remains that immateriality is the key to explaining how cognition can unify the complexity of experience into wholes, which can be experienced in a single, unified act of cognition.

Just as animals and man can do this at our own finite and limited levels, by way of transcendent analogy, the same explanation must be applied to God so as to render intelligible how he can know all things and cause all things, even in their near infinite multiplicity—all the while remaining absolutely simple and undivided in himself. We do not need to know exactly how he does this, any more than we need to know how we do it—in order to know that it is true (1) that it happens and (2) that it can happen solely because of the immateriality of the cognitive powers involved.

The analogy is that just as animals can perceive whole sense objects in a unified way and that man can understand many individual natures in a single concept, so too, God can know all things in a single unified act of understanding which is identical to the divine essence.

And, because of the divine simplicity,[7] since God's act of knowing things is identical with his act of creating them, he both knows and causes to be the innumerable multiplicity of created things in a single, perfectly unified spiritual act.

---

[7] Available online at https://strangenotions.com/divine-simplicity/

Chapter 5

How to Approach the Problem of Evil[1]

The problem of evil in relation to God's goodness is too vast a topic to treat fully in this short article. Therefore, I shall offer just a few relevant observations on this widely known objection to God's goodness and existence.

In classical metaphysics, proving God's goodness starts with defining what is meant by the good. The good is that which all things desire.[2] But a thing is desirable because it is perfect, which implies that it is as actual as its nature permits. Since a thing has being as it has actuality, the good is equivalent to being.[3] Since God is infinite being,[4] he must also be infinite goodness.

Moreover, since all valid proofs for God's existence argue from finite effects to God as the First Cause of all creatures, he must be the cause of the goodness found in all things.[5] Since a cause cannot give what it does not have, God must possess goodness. But the divine simplicity entails that God is identical to any quality he

---

[1] This article first appeared online on the *Strange Notions* website. https://strangenotions.com/how-to-approach-the-problem-of-evil/

[2] Aristotle, *Nicomachean Ethics*, I, 1 (1094a 1).

[3] *Summa Theologiae*, I, q. 5., a. 1, c.

[4] Available online at https://strangenotions.com/how-cosmic-existence-reveals-gods-reality/

[5] Available online at https://strangenotions.com/how-gods-nature-is-known-the-three-fold-way/

possesses.[6] Hence, God is pure goodness. Since moral goodness is a genuine form of goodness, God must possess and, in fact, be moral goodness itself.

Once it is demonstrated that God is morally good, the solution to the problem of evil requires only that one understand how evil can exist in spite of God's goodness. *In other words, since the problem of evil does not arise until we already know that God exists and is infinitely good, it is therefore a given that the problem of evil can be rationally resolved.*

On the other hand, for atheists or agnostics who approach the problem of evil without knowing that God exists, it is the existence and goodness of God that are in jeopardy, since they are certain that evil exists and appears a vexing problem. So, they are in serious doubt that an all-good God can possibly exist.

Clearly, it makes enormous difference as to how one approaches the problem of evil. For the theist, it is merely a problem to be solved. For the atheist, it is a massive obstacle to belief in a good God. It all depends where one starts his enquiry.

Since classical metaphysics *does* demonstrate the existence of an all-good God–and since I have published defenses of such arguments,[7] mine is the former task. It is merely a matter of seeing why the world's evil is compatible with the all-good God already known to exist. From this perspective, atheists and agnostics simply approach the problem from the wrong end.

---

[6] Available online at https://strangenotions.com/divine-simplicity/

[7] Available online at https://www.amazon.com/Aquinas-Proofs-Gods-Existence-Necessarily/dp/940118187X

Chapter 5: How to Approach the Problem of Evil

Since the good is equivalent to being and good and evil are diametrically opposed, it would appear that evil must be simply non-being. But, evil is *not* simply non-being. Rather, *evil is the lack of being or perfection that should belong to a given nature.*[8]

Physical evil is the privation of a natural physical good, as when a horse has a broken or missing leg. For many, evil is viewed as pain and suffering. These, too, represent a lack of well-being in sensation or feeling. Moral evil is the lack of rectitude in the acts of a free agent—a sin.

Why Does God Permit, Or Even Cause, Evil?

It is often argued that,[9] if God is all good, all powerful, and all knowing, he has no excuse even for *permitting* evil to exist. It appears that either he is not all good, or he is powerless to prevent evil, or he does not know what is going on. None of this is compatible with the classical conception of God.

Nonetheless, *it is morally licit to permit evil—when that permission allows a greater good to result.* For instance, I might allow a youngster to smoke a cigar, knowing it will make him sick, but for the greater good of teaching him not to smoke at all. Now, this is not the immoral act of causing an evil means so as to attain a good end, since I am not making the youngster smoke the cigar. *That is his act, not mine.* So, too, since God gave us the perfection of a free

---

[8] *Summa Contra Gentiles*, III, ch. 6, para. 1.
[9] Available online at https://en.wikipedia.org/wiki/Problem_of_evil

will, he can allow us to misuse that will and sin, while knowing and willing that a greater good may be forthcoming.

Since God is infinitely good and powerful, it necessarily follows that any evil that God permits in this world must have a greater good that results from it. Being infinitely powerful and knowing all future events, *God's goodness could not permit that evil should occur unless greater good is foreseen to ensue from it.*[10] The fact that we cannot conceive of such a greater good in many cases does not demonstrate that God is evil, but rather that *our finite minds cannot understand the inscrutable nature of God's providential plans.*[11]

Still, the question arises as to whether God, not only permits evil, but directly causes it in some instances. Clearly, when God exercises his divine prerogative over creatures, as in the matters of life and death and punishment, he acts in ways that entail physical evil for his creatures. How then does God remain free of moral evil when he directly causes such physical evils? Many argue that when God directly takes human life or administers other punishments, he is acting immorally—even manifesting brutality.

But God is the Creator and Sustainer of all life—life, which is given to us as a gratuitous gift. What is freely given may be freely withdrawn at any time—with no resulting injustice. Moreover, *as the divine lawgiver and judge of natural law, God is perfectly right to punish directly its violators—so as to restore the balance of justice.* No mere creature has that prerogative.

---

[10] *Summa Theologiae*, I, q. 2, a. 3, ad. 1.
[11] Karlo Broussard, *Preparing the Way* (Catholic Answers Press, 2018, 79-82).

Now, no one would say that it is illicit to remove surgically a cancerous organ, even though the necessary first step is to cause the physical evil of making an incision, which can be painful and damages skin—for it is clear that the total act involved is that of removing a threat to human life.

So, too, when God imposes licit sanctions on evil men that entail pain and suffering, the total act is that of imposing the sanction or punishment, while the good end or purpose of that act is the restoration of the balance of justice. Sanctions themselves are a social good needed for the upholding of laws. The pain and suffering (or even death) are simply an essential part of imposing the sanction, which cannot be separated from the act itself. The somewhat incidental nature of the form of the sanction is evinced by the fact that differing crimes receive differing penalties, whereas what is constant and common is the concept of the sanction itself.

In any case, it must be emphasized that *God would never will physical evil (either directly or indirectly) for its own sake, that is, as an end in and of itself.* He always wills it within the framework of the good of the whole of the created order. We must also remember that what is morally evil for man may not be morally evil for God, since he alone is the Creator of all things and the Legislator of natural law as well as the just Judge of those who violate its ordinances. For example, humans can never licitly take an innocent life, but God *can* do so—given his position as Creator and Sustainer of all finite living things.

It is self-evident that the infinitely-good God could never directly will moral evil for the sake of any end whatever—however good.

Because some, such as Luther, Ockham, and Descartes, embraced classical positivism with respect to God's will, they thought he could make adultery licit—or even make two plus two equal five. Natural law never allows such absurdity, because God respects the nature of his own plan of creation. Thus, God could never make adultery or *odium dei* licit or, for that matter, make two plus two equal five.

This general explanation dealing with God both permitting and causing evil completely resolves the problem of evil in all its many forms, since whatever evil occurs in the world can only happen because God foresees and wills a greater good coming from it.

This solution follows necessarily from the facts that God's existence can be demonstrated, as can his infinite goodness, power, and knowledge.

But what of the atheist's or agnostic's perspective, since he does not accept these metaphysical conclusions about God and his goodness? Coming from a given starting point of the existence of massive evil in the world, it would seem that the hypothesis of an all-good God is *a priori* excluded.

Quite to the contrary, it is the atheist's or agnostic's burden of proof to show that such evil is incompatible with an all-good God. For, if God does exist as classically depicted, then it follows that the problem of evil dissolves as explained above. For the atheist or agnostic to prevail, he must show that such a good God does not exist. He argues from the existence of evil to his conclusion that an all-good God cannot exist. But that is begging the question, for he is assuming what he purports to prove. As we have already shown above, *if the God of classical tradition does exist, then evil is no problem.*

Thus, the problem of evil is resolved no matter which end of the question is addressed first—be it the existence of evil or the existence of an all-good God.

This means that in principle this analysis and solution of the classical problem of evil could end at this point with no further discussion. Nonetheless, I shall consider some further aspects.

The Problem of Pain

If the problem of evil were a purely rational objection to God, it would seem that every kind of evil should be concerning. Yet, I have never heard anyone proclaim his atheism because of the carnage taking place against lettuce when a chef prepares a salad. Still, there is concern about the pain and suffering that animals endure. Human animals are well aware of the agony that pain can cause. Even so, concern is selective. I have never heard anyone proclaim his atheism because of the treatment of bugs in a Raid commercial.

Animals naturally experience sense pain and pleasure. The sense appetites move them to seek the pleasurable good. They also move them to avoid sensible evil: displeasure or pain. Animals seek goods that keep the individual alive and the species reproducing. *Animals need to experience and to fear pain in order to survive against threats to their lives and those of their offspring.*

One might ask why God didn't make animals so that they did not experience pain. The answer is that, in this natural world, pain plays so central a role in animal life that the only way to avoid the problem of pain for animals would be to eliminate the animals themselves. But this is absurd, since (1) it would limit God's power to create life and (2) it would solve a problem by eliminating the very beings it seeks to benefit. Better for animals is that they live with some pain at times, rather than not live at all.

On the other hand, pain in human experience must be considered in the broader context of man's intellectual and spiritual life and its role in helping him attain his last end.

Evil as Part of God's Plan for Man

The problem of physical evil and pain in human existence must be subordinated to a proper understanding of his last end and the role of free will in his attaining that end.

When we look at this world, so filled with evil and suffering, the question naturally arises, "How could a good God make such a world?" But this presumes that God is totally responsible for the world as it *now* exists. Perhaps, God made a world without evil, but he also created free beings who made evil choices that might have corrupted all creation. If evil's existence before man's coming be objected, one must then consider the possibility that God created other free beings, such as angels, prior to human creation, and those free beings introduced evil into the world.

Other possibilities include: (1) that the reward of heaven might not be justly given without man earning it. (2) that an *earned* reward is more perfect than an unearned one. (3) that pain and suffering are key elements in progress toward moral perfection.

God *could* have made his own existence so evident that no free creature would *dare* misuse his freedom, and thereby, fail to attain his last end: heaven. Instead, God created an evolving natural world that permits the possibility of naturalistic explanations for everything. In a word, God made a world perfect for atheists and agnostics—since they can argue plausibly (but, not correctly!) that God is a useless myth.

We live on a wonderful planet that rotates so that human life is possible, but whose resulting weather patterns cause death-dealing hurricanes. Humans thrive, but physical evils abound. Could God provide countless miracles to save endangered lives? Could God have made the cosmos differently? Perhaps. But, would the world still be best suited to allow *maximum human freedom* in reaching our last end?

While right reason can lead the human intellect to affirm God's existence, man's spiritual destiny, and the force of natural law, no one is virtually coerced into this awareness as he would be if God's existence were undeniably evident. Unfortunately, this scenario also entails the possibility of man readily misusing his free will so as not to attain his last end. Why would God permit such a self-destructive use of human freedom?

We might prefer "forced" salvation, but God respects his creature's freedom so much that he allows us *meaningfully* to freely earn our eternal reward—even at the price of possible *deserved* failure. *A free agent's greatest qualitative perfection is most perfectly achieved when he freely chooses a life of moral virtue, even when aan evil alternative deceptively beckons—as in the mod-*

ern secular world, which seems to offer paradise on earth with no difficult moral constraints, such as sexual self-control.

This world, in which evolutionary naturalism appears to be a real alternative to God's presence and plan, turns out to be the perfect world for the building of the greatest of saints.[12] This world necessarily entails the presence of great evils—the worst of them being of human making. Still, the fact remains that God has good reason to create this world exactly as he has, since its *evils exist only because God foresees far greater good forthcoming as a result, that is, a heaven filled with creatures who freely merited their eternal reward.*

---

[12] My final argument showing that a naturalistic world is best designed for maximum freedom is taken from my book, *Origin of the Human Species–Third Edition* (Sapientia Press, 2014), 211-213.

Chapter 6
Hell and God's Goodness[1]

Although this article will address the content of certain theological doctrines, it is written from a purely philosophical perspective. This is the same method used consistently in my book, *Origin of the Human Species*,[2] in which I examine how evolutionary theory comports with divine revelation and philosophy. What characterizes philosophical analysis of theological doctrine is that reason alone is the method employed. Thus, while the philosopher as such cannot say whether the Trinity is factual, he can still examine whether it appears rationally possible.

The Scandalous Problem

Here I will examine the theological doctrine of hell to see whether it is compatible with the God of classical theism, who is claimed to be all good, all loving, and all merciful.

Many skeptics seem to think it is obvious that an all good and loving God could not possibly consign a fallible human being to the unimaginable, interminable, excruciating pain of physical fire and other torments in the form of punishment known as hell—a sanction for sin from which there is no appeal and no hope of fu-

---

[1] This article first appeared online on the *Strange Notions* website. https://strangenotions.com/hell-and-gods-goodness/

[2] Available online at https://www.amazon.com/Origin-Human-Species-Dennis-Bonnette/dp/1932589686

ture release. Surely, no good God and no compassionate human being could possibly even contemplate such unmerciful treatment of a human soul, merely because she made errors of choice during a single short lifetime.

One Aspect of the Solution

I do not intend to address every possible solution to this specific variation of the well-known problem of evil–a topic I have dealt with in more general terms elsewhere.[3] Among possible solutions, it is argued that God permits evils, including physical suffering, for some greater good, which the human mind cannot grasp. Since metaphysics proves that God is all good, it necessarily follows that any evil found in the world cannot be his fault. Perhaps, the misuse of free will by certain creatures (angelic or human) has led to the introduction of evils unintended by God. Perhaps, man's misuse of free will calls forth from divine retributive justice a punishment which seems severe, but which must be measured in terms of the infinite goodness which grievous sins offend–thus requiring the eternal pains of hell as a just punishment.

Pointedly, since God is the transcendent Law Giver, he is not bound by the natural laws that apply to creatures. Rather, it belongs exclusively to him to administer retributive justice to those who violate his laws—natural and divine. This means that it is good that

---

[3] Available online at https://strangenotions.com/how-to-approach-the-problem-of-evil/

God punish the wicked as part of his overall plan of creating and governing a good and just world.

But Why the Pains of Hell?

Still, I focus here on the specific question posed by some skeptics as to why an all-good God would submit departed souls to eternal physical pain and suffering, even in its most agonizing form of physical fire?

Certainly, it appears at first glance that such suffering is nothing but an act of pure vengeance on the part of God. Indeed, does not Scripture declare, "*Revenge is mine, I will repay*, saith the Lord"?[4] But, how can revenge be reconciled with the concept of a loving God?

Yet, while revenge is not an act permitted to mere mortals, it does have a legitimate meaning properly reserved to God as the ultimate administrator of retributive justice. Retributive justice is not just "getting back" at someone, but the restoration of the proper order of things—an order in which each person gets exactly what he deserves, including proper punishment for his evil deeds. Moreover, it must be understood that this right belongs in its highest instance to God alone as creator and supreme lawgiver.

Should such retributive justice include the fire of hell? And, if so, how can this be reconciled with the belief that God is all good and loving and merciful?

---

[4] Romans 12:19 (Douay-Rheims).

The Specific Solution

Most skeptics' accusations against the punishments of hell are made on the supposition that even the Christian understanding of creation does not justify such eternal sufferings.

I will show that this divine retribution is consistent with the general order of creation presented in Christian sources as well as with the infinite goodness, justice, and mercy of the God of classical theism.

Among the central doctrines of Christianity is that man's last end—the ultimate purpose of his very being—is to be united with God for all eternity in a face to face encounter with the divine being, what Catholics call the Beatific Vision.

But, in this present life, we do not enjoy the Beatific Vision. Does that mean that we are already in hell? Well, yes and no. Yes, in the sense that we are presently lacking the ultimate end of God's intention in creating our nature. But, more importantly, we are not now in hell, in the sense that this is neither a punishment nor necessarily an eternal condition.

The essential meaning of hell is (1) that we finally understand fully that the Beatific Vision is the sole reason for our creation as human beings and the sole thing fully worth accomplishing in our existence, (2) to know that this end will for all eternity be denied to us, and (3) to know that this ultimate failure of our existential purpose is totally and completely our own fault and no one else's.

All this being the case, why does not God simply punish bad lives by letting us merely fail to accomplish our intended final bliss in exactly the manner just described? Why is the threat of extreme physical punishment seemingly arbitrarily and capriciously attached to this natural spiritual sanction for a wicked life?

Man is a Rational Animal

Man's uniqueness is that his spiritual, rational soul is embodied in an animal nature. Being an animal means that our nature is that of a sentient organism. We have senses. Our sensitive appetites help us achieve the good of the individual and of the species by making us seek sensible goods and avoid sensible evils. The natural good that accompanies attaining the sensible good is pleasure. The natural evil that accompanies experiencing sensible evil is pain. Thus, we are strongly motivated to seek pleasure and avoid pain.

It is not merely man's spiritual soul that is aimed at his last end, but rather it is his whole human nature—spirit and *matter*, soul and *body*—that either attains his last end or fails to attain it through his own fault.

For this reason, we should not be surprised to discover an important role is played by bodily sense knowledge with respect to how and whether man reaches his last end, the Vision of God.

If we believed that missing our last end meant merely never seeing God in his very essence–never having the Beatific Vision, that might not be a sufficient reason for many people to lead virtuous lives. Such a "purely spiritual" motivation might not move us the way that we, like other animals, can be intensely motivated by desire for pleasure and, even more so, fear of agonizing pain!

Many would say that they do not presently miss God's presence all that much in this life anyway, so why worry about missing him permanently in the next? In truth, we are *not* moved decisively in this present life toward God in all our choices—despite many of us knowing, in theory at least, that he is the highest good.

In a word, to many people in this life, if all they thought the end of life entailed was attainment of the Vision of God, they might well be inclined to forgo that final destiny, since they would easily not value it as much as the earthly pleasures they know would never lead them to such alleged heavenly bliss!

But God made man to fear physical pain—and properly so, since it moves us to avoid dangers to our well-being both as individuals and for the sake of our species' survival.

Therefore, it makes eminent sense that God would use man's intense fear of great pain to motivate him to reach his last end. Once in the next life, man will clearly know the value of the spiritual reward of the Beatific Vision. Those who fail to attain that true last end through their own fault will then have the appalling realization of failing to attain the very purpose of their existence. But, in this life, the intensity of most human beings' motivation is focused on sensible rewards and punishments, on pleasure and pain.

Thus, the realistic possibility of knowing that we may fall through grave sin into an eternal pit of most intense physical pain would be, for most mortal men, the strongest possible motivation to live an essentially good life—a life best ordered to avoiding the physical suffering of eternal damnation.

Chapter 6: Hell and God's Goodness

God's Love

Is letting sinners go to hell then truly an act of simple retributive justice on the part of God? Is God seeking merely to punish their moral evil by allowing them to fall into the pits of hell?

On the contrary, *it is the greatest act of love on God's part to make certain that men are motivated as strongly as possible to seek and attain what is, in truth and in fact, the greatest possible happiness—the eternal Beatific Vision.*

In other words, we humans do not properly value what will make us happiest in the long run, and thus, through our own craven ignorance of proper goods, fail to attain the perfect bliss God wills for all men in his act of creating them. Hence, God makes certain that we are properly motivated to seek our true and most perfect end, by graphically placing before us the sensible horror that confronts those who willfully fail to attain their proper last end.

But Most People Don't Even Believe in Hell!

That is quite true. And, even among those who should do so—based on their religion's public doctrines, a large number do not believe in hell's physical reality. Of the roughly 7.6 billion people on Earth as of 2018, about 55% belong to Judaism, Christianity, or Islam—all of which religions have some real notion of hell. That makes for some four billion people. If even half of this total take the torments of hell seriously, as probably do, that makes a total of about two billion people–or roughly a quarter of the world's population—that believes in the real pains of hell.

Therefore, a good portion of humanity is motivated by the physical pains of hell to seek salvation seriously. *Fear of hell can well be the beginning motivation that leads one to those religious practices, which, in turn, may lead to a more mature and deeper appreciation that one's highest motivation should be, not fear of hell, but love of God because of his infinite goodness and perfection.*

St. Thomas makes much the same point: "From becoming accustomed to avoid evil and fulfill what is good, on account of the fear of punishment, one is sometimes led on to do so likewise, with delight and of one's own accord. Accordingly, the law, even through punishing, leads men to being good." (*Summa Theologiae*, I-II, q. 92, a. 2, ad 4.)[5]

Thus, the fear of hell may set one on a path leading to virtuous living for its own sake. This, in turn, can lead to the true understanding that union with God is the essence of heaven and our proper last end.

While eternal life for the inhabitants of paradise in Islam is usually associated with sensual pleasures, Islamic teaching also affirms that "the most acceptable of them with God shall look upon His face night and morning."[6] (Al-Qiyama 75:22,23)

---

[5] "Ad quartum dicendum quod per hoc quod aliquis incipit assuefieri ad vitandum mala et ad implendum bona propter metum poenae, perducitur quandoque ad hoc quod delectabiliter et ex propria voluntate hoc faciat. Et secundum hoc, lex etiam puniendo perducit ad hoc quod homines sint boni." Editio Leonina.

[6] http://www.message4muslims.org.uk/muslim-doctrines/qa-the-last-things-eschatology/heaven-hell-islam/

In other words, for at least a quarter of mankind, belief in a literal physical hell serves the purpose of leading men in the direction of virtuous living. The net effect of this motivation toward salvation would naturally also lead many to understand the true value of our last end as being the Vision of God. From this would naturally also result an increasing number of souls seeking to please God by living more and more holy lives, that is, to achieve genuine sanctity.

Thus, if one wonders why God would make the torments of hell central to the beliefs of what is, *de facto*, only, perhaps, a quarter of mankind today, the answer might just be that God is actually concerned with the spiritual *quality* of human perfection. That is, *God may be concerned not merely with the quantity of the saved, but also with the qualitative perfection to be found among those who are saved.*

Oddly enough, while beliefs about the eternal torments of hell are used by skeptics as reason to disbelieve in God's goodness, those same beliefs may motivate far greater numbers of souls to follow an upward journey of religious understanding that leads them eventually to the most holy religious insights and practice, that is, to sanctity itself.

In a word, the doctrine of hell creates a world designed to produce the greatest of saints—*a qualitatively more perfect end than might otherwise be possible without hell.* It is perfectly within the prerogative of God to design his creation so as to produce the most spiritually perfect creatures. Since the fear of hell, *a hell which is licit in itself as a form of divine retributive justice,* can serve as a licit means to that more perfect end, it is fully justified.

The Doctrine of Hell and Free Will

The Catholic Church dogmatically teaches that those who die in a personal grievous sin descend immediately into hell[7] and that the punishment of hell lasts for all eternity.[8] Nonetheless, while the majority of traditional theologians do believe hell to entail a physical fire, it also remains true that the Church has never condemned the speculation that hell's "fire" is constituted of purely spiritual pain, such as exclusion from the Beatific Vision and the pangs of conscience.[9]

Moreover, the *Catechism of the Catholic Church* declares that this "exclusion" from the Beatific Vision is essentially a form of "self-exclusion."[10] Such self-exclusion is expressed by Lucifer in Milton's *Paradise Lost*, when he proclaims, "Better to reign in hell than to serve in heaven." It is a measure of final obstinacy and pride that refuses to abandon serious sin and accept divine forgiveness.

It is true that the rational appetite, or will, must always choose the good. But does that mean that no free person in full possession of his faculties could refuse the highest good, God himself, so as to "self-exclude" himself from heaven and go to hell as a result?

---

[7] Solemn declaration by Pope Benedict XII in the Dogmatic Constitution, "*Benedictus Deus.*" Denz. 531.

[8] The *Caput Firmiter* of the Fourth Lateran Council. Denz. 429.

[9] Ludwig Ott, *Fundamentals of Catholic Dogma*–sixth edition (B. Herder Book Company, 1964), 480-481.

[10] Catechism of the Catholic Church, 1033.

This question reveals a basic misunderstanding of the nature of the rational appetite or free will. The good is defined as being as desirable. So, the rational appetite naturally desires every possible good. That is to say, the will is necessitated to seek the universal good or happiness. But the human will does not desire any particular good—no concrete good or action—necessarily in this life, since particular goods are good under one aspect, but not under another.[11]

Finite goods can always be refused, since there are elements of imperfection about them which may be possessed by some other goods. Hence, we choose between various goods, like chocolates in a box, where each has qualities lacked by others and vice versa—thereby, forcing us to choose between them.

But with regard to direct knowledge of God in his essence, St. Thomas Aquinas concludes that "the will of him who sees God in his essence of necessity adheres to God, just as now we desire of necessity to be happy."[12] (*Summa Theologiae*, I, q. 82, a. 2, c.)

God is known in his essence solely in the Beatific Vision. Yet, it is self-evident that the final refusal or acceptance of God must take place *before* God is embraced fully or rejected completely.

Hence, *God is not known in his very essence before the soul reaches heaven.* What the soul knows before that time must then be some finite good, such as knowing the truth that God is the highest good. *But, any finite good can be refused.*

---

[11] Bro. Benignus Gerrity, *Nature, Knowledge, and God* (Bruce Publishing Company, 1947), 244-245.

[12] "Sed voluntas videntis Deum per essentiam, de necessitate inhaeret Deo, sicut nunc ex necessitate volumus esse beati." Editio Leonina.

The problem is that we often choose lesser and improper goods even when we know that they are opposed to God's law or to God himself–or *even to our own true good!* In fact, the very basis of our experience of free will is the fact that we can choose between various finite goods and even choose sinful goods that we know are opposed to the true good, or even the goodness of God himself!

Unfortunately, this is precisely why a hardened sinner, who still has essential possession of his rational faculties, can freely exclude himself from heaven by stubbornly rejecting the law of God or even some finite representation of divine majesty and love—even on his deathbed.

How Many Are Lost?

But how many people actually go to hell? On that question, the closest the official Magisterium comes to offering an answer is found in the encyclical by Pope Benedict XVI, *Spe Salvi*, in which he suggests that, while a few souls go directly to heaven and a few go directly to hell, the "great majority of people" go to a place of temporary purification before entering the Vision of God, the place Catholics call "purgatory."[13]

Curiously, even Islamic writings seem to have some notion of limited duration of punishment in hell, a concept similar to the Catholic conception of purgatory. Indeed, <u>one optimistic text</u> proclaims that almost all people will be removed from this

---

[13] *Spe Salvi* (2007), nn. 45-46.

state of suffering: "From every one thousand, take out nine-hundred-and ninety-nine." (Bukhari 4:567)

Since there is no way to be certain just how many, if any, souls actually go to hell, it is also possible that those who—*through no fault of their own*—are ignorant of its existence, may find themselves more likely to end up in some form of purgatory, rather than hell itself.

Conclusion

On careful reflection, the notion of hell as a place of eternal punishment for the souls of the wicked after death turns out to be (1) a just application of retributive justice by a Divine Lawgiver who stands ontologically above the natural law of his creation, (2) a natural sanction that is actually self-imposed by a will stubbornly opposed to the righteous laws of Infinite Goodness, and (3) a powerful tool designed to use the natural avoidance of pain—both spiritual and physical—as a motive to follow God's laws and prepare souls for a spiritual ascendancy leading to the direct vision of God himself, which is man's perfect happiness.

Hell, then, is not something evil in itself, but a natural byproduct of the order of being, one which aids in bringing many human beings to the highest state of natural—or even supernatural—perfection through holiness of life. God's divine providence aims to produce the happiest creatures possible.

Chapter 7

Theism vs. Skepticism: The COVID-19 Pandemic[1]

In an earlier *Strange Notions* essay,[2] I addressed the problem of how an all-good God could be compatible with the existence of Hell. While that analysis befits the extreme case, the purpose of the present piece is to address the exact role of responsibility God has in terms of the very real and human tragedy posed by the Covid-19 virus which is presently raging throughout the world.

This piece will not address the most ethical or medically correct methods with which to address this pandemic. Rather, its sole purpose is to understand the role that God plays in allowing, supporting, and/or causing Covid-19's enormous toll of pain and anguish on mankind.

---

[1] This article first appeared online on the *Strange Notions* website. https://strangenotions.com/theism-vs-skepticism-the-covid-19-pendemic/ It was written during the initial phase of the Covid-19 illness, before it was realized that the true infection mortality rate was 0.5% or less, with the corrected median rate for those under age seventy being only 0.05%. Its logic, though, could apply to "any other great evil." Cf. Aloannidis, J.P. (2020, Oct. 14). Infection fatality rate of COVID-19 inferred from seroprevalence data. National Library of Medicine. Retrieved November 30, 2022, from https://pubmed.ncbi.nlm.nih.gov/33716331/.

[2] Available online at https://strangenotions.com/hell-and-gods-goodness/

Classical Theism's Defense of God's Goodness

In defense of God's goodness, classical theists will point out that God is not a moral agent as mere creatures, such as men and angels are. As Creator, he is free to take back the gift of life he has given to men. Yet, it is also possible that the physical and moral evils we see in the world are caused by the actions of free creatures. Since God has chosen to create creatures with intellectual natures, both angelic and human, such beings are inherently free.[3] With freedom comes the possibility of deliberate fault or sin. And thus, what God has created with perfection in the first place may become corrupted by free creatures' misuse of their freedom. Such theological doctrines as original sin describe how free agents, such as human beings, could introduce real evil into a world originally created by God as good.

Evil that appears in the world may be (1) the result of a free agent's misuse of freedom in a particular act that results in both his own corruption and evil effects that are of his making, (2) the result of some kind of primeval fall by an angelic order of beings that infected the rest of subsequent creation, or (3) the product of a human original sin that perverted the natural goodness of later men and the order of nature itself. Some such scenario could be responsible for such natural physical evils as the Covid-19 pandemic together with all its suffering and death.

---

[3] *Summa Theologiae*, I-II, q. 6, a. 2, ad. 2.

Chapter 7: Theism vs. Skepticism: The COVID-19 Pandemic

But skeptics rightly probe more deeply and ask precisely how God can be the ultimate cause of all things, and yet, claim no moral responsibility for something as horrific as the Covid-19 pandemic?

Thomists typically explain that creatures have genuine secondary causality, whereby their actions are properly their own, even though God sustains them in their execution. Thus, evil is introduced to the world either (1) through chance interactions of secondary causes, or else, (2) through the free agency of either angelic beings or men. In either case, the impression is given that evil's responsibility is assigned to the secondary causes and not to God himself.

But is this the complete story? Surely, many natural agents appear to act to achieve something good for themselves. For example, a lion, seeking to eat, may interact with a gazelle, seeking to drink at an oasis, in a way much to the disadvantage of the gazelle. And, while neither lion nor gazelle is seeking anything evil, evil accrues to the gazelle as a result of their "chance" interaction. "Chance" events, in Aristotelian philosophy, do not mean events with no causation whatever, but something that happens outside the natural tendency of a given agent. Thus, while the gazelle goes to the oasis for water, its chance crossing of paths with the lion results in an unwanted outcome, namely, being eaten by the lion.

The bottom line of such causal confluence is that each agent, while acting so as to produce its own natural results (or, what Thomists argue are perfective ends), may well interact with other natural agents so as to produce an outcome outside the natural tendency of one, or both, agents involved.

Similarly, moral evils committed by free primeval angelic spirits and/or first true human beings might have introduced original disorder into creation, thereby explaining resultant cataclysmic physical and moral evils. While God is responsible for creating the perfection of such free agents in the first place, he is viewed neither as responsible for their misuse of freedom nor for the evil effects resulting therefrom.

But, do these typical explanations really entail that God in no way causes the evil we find in creation, especially as witnessed in a malevolent pandemic such as Covid-19? Quite to the contrary, God's hand remains in every last detail of creation as is clear from the 1913 *Catholic Enclycopedia* explanation of Divine Providence:[4]

"God preserves the universe in being; He acts in and with every creature in each and all its activities. In spite of sin, which is due to the willful perversion of human liberty, acting with the concurrence, but contrary to the purpose and intention of God and in spite of evil which is the consequence of sin, He directs all, even evil and sin itself, to the final end for which the universe was created."

God not only causes the very being of all creation, but he keeps every particle of it in existence at all times. Moreover, as it changes and undergoes motion, God is the cause of the very existence of all that comes-to-be as new in finite reality.[5]

---

[4] Available online at https://en.wikisource.org/wiki/Catholic_Encyclopedia_(1913)/Divine_Providence

[5] Available online at https://strangenotions.com/how-new-existence-implies-god/

This means that, while creatures, acting as secondary causes, are true causes of their own actions, such actions could never take place without God (1) sustaining the being of those agents and (2) also acting as the ultimate cause of every new quality of being that results from their actions.

The Nature and Role of Chance Events

As for chance events explaining evil in the world, many people do not realize that chance has two meanings: (1) an event taking place somehow spontaneously without any real cause, and (2) the classical Aristotelian notion of chance described above as something happening outside the natural tendency or intention of an agent.

Today, many people think of chance events as things happening without any real cause. Specifically, some interpret Heisenberg's Indeterminacy Principle as meaning that there are subatomic events whose manifestation is not dictated by any actual cause.[6] Other leading physicists, including Schrödinger and Einstein, maintained that this renunciation of deterministic causality was physically incomplete. Far more importantly, this denial of causality at the subatomic level is metaphysically impossible, since that would amount to having being come-to-be from non-being.

Metaphysically, if "chance" means something happening without an actual cause, then there are no such "chance events" at all.

---

[6] Available online at https://www.britannica.com/biography/Werner-Heisenberg#ref524688

The other meaning of chance (described earlier) is philosophically tenable, since it merely refers to something interfering with an agent's movement toward an expected outcome, whether the agent is intelligent or not. For example, one goes to the bank to make a deposit and accidentally meets a creditor who instead demands the money. Such an encounter of diverse causal orders would be called a chance event, but one whose outcome would in no way escape predictability to someone knowing the paths and intentions of both parties.

Similarly, a rock rolling down a hill encountering another rock that blocks its expected path would also be called a chance event, even though the outcome is perfectly deterministic in nature.

From the above, it should be clear that neither type of event called "chance" escapes the foreknowledge and will of God as described in Divine Providence, since (1) "chance events," understood as being purely spontaneous or acausal simply do not exist and (2) God knows the tendencies and interactions of all agents. And, since all natural agents conform to the will of God in determining the course of causal events in creation, it is clear that God would be responsible for the course and outcome of all events, whether called "by chance" or not—barring, of course, interference by free creatures.

Still, Why Does God Enable Free Agents to Choose Evil?

Since most authors realize that the world as understood by classical metaphysics would flow deterministically from God if no free agents existed, the central thrust of explanations of evil focuses on the existence of such free beings. If free beings are really free, then it must be possible that they misuse their freedom, and thus, could introduce moral and physical evil into the world. From that initial appearance of free deviation from God's plan of creation could then be explained the presence of subsequent physical and moral evils, whether they flow directly from evil choices, or, in some hypotheses, even by some sort of temporally antecedent effects anticipated by God's eternal vision.

Therefore, while *God does not directly cause such great evils as the Covid-19 pandemic,* his creation of free beings—angelic or human—*might* explain how such evils come to be without having to blame God himself for consenting to these evils.

As noted earlier, the problem remains that no creature—not even a free one—can perform any act, whether it is viewed as secondary causality or not, without God sustaining its nature and enabling its activity. Thus, while God may not consent to or affirm the freely chosen evil intention of a free agent, he nonetheless sustains the activity of all the physical powers and actions by which an evil deed is performed. He may not will that the evildoer do evil, but he does permit and support all the physical powers by which the evil deed is committed—and even sustains the power of choice of the free agent in committing the evil deed.

I do not intend to argue here whether human freedom is possible, since that is a distinct issue which I have addressed elsewhere.[7] *The question at hand is why does God allow and support*

---

[7] Available online at https://strangenotions.com/how-human-free-will-harmonizes-with-sufficient-reason/

such evil choices and how is he not therefore responsible for their evil? And this is especially problematic in the case of explaining the connection between evil choices and the appearance of a blind, non-living, demonic virus, such as now plagues humanity.

Various hypotheses have been offered as to how creatures' free choices might have resulted in evil entering the world. Theologically, Christians consider the possible effects of Lucifer's rebellion or Adam's original sin. Like a symphony orchestra whose conductor permits a small section to continue playing off tune, eventually the entire enterprise may go off tune—and, perhaps, there is a similar progressive cascade of moral evil precipitating ever greater physical and moral evils in the created world.

Even without some free creatures' initial misdeed, perhaps, God created a world in which cosmic and biological evolutionary scenarios entail such "chance" interactions (in the Aristotelian sense described above) that physical evils result, as in the case of the lion surviving by eating a gazelle. In more dramatic terms, might God have planned a world in which earthquakes, volcanoes, and tsunamis would occur—or even a Covid-19 virus would evolve—for the greater good of reminding mankind that life is short and he still has need for his Creator?

In fact, this world might actually have been planned by God so as to enhance human freedom by making naturalistic evolution a plausible hypothesis for atheists who prefer not to believe in him!

"Since naturalistic evolutionism is the near-universal refuge of atheism, an evolutionary world becomes a world where persons experience maximum moral freedom, including freedom to deny God's existence and moral law."[8]

God may have many reasons to permit free choices which lead to real physical and moral evils. The principle here is that while an evil means is never permitted to attain a good end, nonetheless it is licit to *permit* evil to occur so that a good end is attained, *provided one does not directly promote the evil means*. This is like a father letting his young son smoke a cigar, not because he wishes him to smoke cigars, but because he knows that if the son gets really sick from smoking *this* cigar, he may learn not to smoke them in the future.

The key here is whether the father has the son's true interest at heart and, analogously, whether God's Providence always permits the introduction of evil into creation so as to attain some greater good as a result, for example, by creating conditions conducive to the raising up of the greatest saints, as I have suggested elsewhere:

"Free agents' greatest qualitative perfection manifests when they choose moral good while self-deceptive evil beckons. Naturalism's possibility, the unintended side effect of creature' maximum secondary causality, offers illusory emancipation from moral constraint."[9]

Thus, those who resist this atheistic self-deception and accept moral constraint can achieve a higher sanctity than if God's exist-

---

[8] Dennis Bonnette, *Origin of the Human Species*–Third Edition (Sapientia Press, 2014), 212.

[9] Ibid., 213.

ence was so manifest as to nearly force puppet-like obedience to the wisdom and justice of his commandments.

Why COVID-19 Does Not Tell Us Whether God Exists

The existence of a global pandemic, such as Covid-19, does not, in itself, determine whether such worldwide suffering and death proves or disproves God's existence. This fact should be the key take away from this essay.

The key is to understand that an all-good and all-knowing God *could* have sufficient reason to permit the existence of such a grave evil as Covid-19—so that some greater good might be obtained.

Why, then, is this not a sufficient explanation of the presence of Covid-19 in the world?

One must first grasp that to the agnostic, atheist, or skeptic the existence of an all-good, all-knowing God may simply not be viewed as a real rational possibility. I say this not to challenge such persons' individual reasons for their rejection of all proofs for God's existence. Rather, I am simply pointing out that the reason they see Covid-19 or any other massive form of human suffering as incompatible with the God of classical theism is not so much because of the inherent horror of the evil itself as it is because of the conviction that no God exists whose nature could possibly justify such evils. That is, in their worldview, there simply is no credible proof that an infinitely good and provident God is real.

So, how could there be any rational justification for Covid-19–not to mention Hell?

Conversely, the classical theist, who is convinced that the one, true God exists and that he is all-good and all-wise, can easily conceive that Divine Providence can know and will an end so good as to justify permitting the existence of virtually any evil imaginable. For, theists take seriously the infinity of God in every respect, and hence, would not dare to think that our finite knowledge of the situation can trump the knowledge and benevolence of what God intends.

That is why the question of whether one views Covid-19 or any other great evil as determinative of God's goodness and power and knowledge depends, not so much on the nature of the finite evil at issue—not even of Hell itself,[10] but upon one's prior intellectual commitment as to whether or not God actually exists and whether he possesses the infinite perfections and attributes ascribed to him by traditional metaphysics.

In a word, I think that the fundamental distance between the way skeptics and theists look at reality as a whole helps explain why unbelievers see the Covid-19 pandemic as just one more proof that God does not exist, whereas believers understand that an all-loving God is reminding us that life in our modern technological age remains radically contingent and desperately in need of its transcendent Creator.

---

[10] Available online at https://strangenotions.com/hell-and-gods-goodness/

Chapter 8

Why Natural Law Ethics is Rational[1]

This article will lay out the rational foundations of natural law ethics as well as show how they lead to implications for the philosophical science of ethics.

Everyone thinks he is an expert in ethics—or, so it seems. Just ask anyone's opinion about any hot topic, like abortion, the homosexual agenda, the proper response to climate change, or the death penalty, and you will elicit strong judgments as to what we are obliged to do or not do. Rarely does someone say, "Who am I to judge?"

Still, depending on one's world view, very different responses to such issues tend to follow. Those who hold that the God of classical theism exists and that man has a spiritual and immortal soul tend to hold radically diverse views from atheists, who hold man is simply the end product of material evolution.

That is why arguments about the ethical inferences of such divergent views tend, at best, to result in a respectful agreement to disagree.

It goes without saying that, if the God of classical theism does not exist, or if there is no unchanging essential human nature or spiritual afterlife, then natural law ethics is mere fantasy. On the

---

[1] This article appeared first online on the *Strange Notions* website. https://strangenotions.com/why-natural-law-ethics-is-rational/

other hand, mankind's universal sense of conscience and compulsion to do good and avoid evil, combined with recognition that some acts are so heinous that history itself offers universal condemnation, such as the Holocaust, make some purely evolutionary explanations appear superficial.

Nonetheless, having already published articles on *Strange Notions*[2] that demonstrate the existence of God[3] and the immortality of the human soul,[4] I hope now to show how natural law ethics, as taught by St. Thomas Aquinas, can be successfully built upon such a body of truths.

Given its necessarily short treatment here, this piece is, at best, but a scanty outline of natural law's rational basis and essential structure.

Eternal and Natural Law

St. Thomas Aquinas defines law in general as "… an ordinance of reason for the common good, promulgated by him who has care of the community."[5] Since he maintains that "… a law is something pertaining to reason,"[6] the natural law always pertains to the "order of reason." It is *not* based on supernatural revelation.

---

[2] Available online at https://strangenotions.com/how-cosmic-existence-reveals-gods-reality/

[3] Available online at https://strangenotions.com/how-new-existence-implies-god/

[4] Available online at https://strangenotions.com/how-we-know-the-human-soul-is-immortal/

[5] *Summa Theologiae*, I-II, q. 90, a. 4, c.

[6] *Summa Theologiae*, I-II, q. 91, a. 2, ad 3.

Chapter 8: Why Natural Law Ethics is Rational

Since God has care of all creation, St. Thomas's definition of eternal law is "... the plan of divine wisdom in as much as it is directive of all acts and motions."[7] God implements his eternally-known plan for creatures quite naturally, that is, through the natures of things themselves. For, what is more natural to a thing than to act in accordance with its own nature?

For physical things, God is the supreme designer of the natural physical law. Physical creatures follow God's directive insofar as each thing must operate according to its own physical nature. Thus sodium must act as its nature directs, when it combines with chlorine to form salt. No one would suggest that a natural body could somehow choose to ignore its own nature and behave like something else. So, too, is the case with all living creatures less than human beings, since, lacking free wills, their behavior is determined completely by their natures–and their natures are the result of God's eternal plan of creation.

Thus, the central insight of natural law is that it operates in and through a creature's very nature, where the nature is the essence of the thing viewed from the perspective of what governs all its activities. Natural law is promulgated in virtue of its being the very principle of operation in every creature. It need not be known by the creature, since natural law automatically dictates how the creature exists and operates. Still, while non-rational creatures do share natural law in a secondary sense, natural law is primarily understood as the *rational creature's participation in the eternal law.*

---

[7] *Summa Theologiae*, I–II, q. 93, a. 1, c.

Man's Last End: Union with God

When it comes to the rational creature, which is man, his possession of reason and free will distinguishes his participation in natural law from that of non-intelligent creatures. This does not mean that human beings' non-free activities are so distinguished, but that the concept of specifically human acts is restricted to those over which we have deliberate control.

Thus, our bodies are subjected to the same physical law of gravity as all non-rational bodies, so that falling from a height would be considered the act of a human, but *not a specifically human act*. So, too, is the case with our physically-determined biological processes. But, a deliberate choice, say, to accept a bribe—since it is under rational and free control, would be viewed as *a specifically human act*.

For the science of ethics, human natural law pertains solely to such freely chosen actions.

Since there is no room here for a complete treatise on natural law, I will limit my remarks to those most pertinent to man's ethical situation. In fact, it does not concern us whether non-human agents universally exhibit final causality, since what is evident is that humans exhibit direct and deliberate actions for ends they understand and freely choose.

We actively seek ends we find fitting to our natural desires, which is why we call them "good."[8] St. Thomas spends the first sixty-three chapters of Book III of the *Summa Contra Gentiles* show-

---

[8] *Contra Gentiles*, III, c. 3, n. 3.

ing that no finite good can completely satisfy the human appetite, since (1) we can always conceive a more perfect good, and (2) even the most perfect goods in this life will be lost at death. Man is never completely happy until and unless (1) there is no greater good to be attained, and (2) the good attained can never be taken away from him.

Since philosophical psychology demonstrates that man has a spiritual and immortal soul,[9] St. Thomas reasons that man's true end cannot be fulfilled in this life, but only in the afterlife. Moreover, since all good things come from the Creator, God must not only have goodness as the cause of goodness in creatures, but, in light of the divine simplicity,[10] must be Goodness itself. Since man is never satisfied as long as a greater good can be had, man's last end must be God himself, who alone is infinite goodness.

From this, it follows that all man's free actions should be directed toward, or certainly not opposed to, attaining God as his last end in the unending afterlife. Since the omnibenevolent God has given man his rational nature and last end, God's own fidelity assures man that proper use of that rational nature will enable us to attain our last end, which is eternal union with the Supreme Good, which is God himself.

The Basis for Moral Obligation

But, man's essence is to be a rational animal. Man differs from, and

---

[9] Available online at https://strangenotions.com/how-we-know-the-human-soul-is-immortal/

[10] Available online at https://strangenotions.com/divine-simplicity/

is superior to, lower animals by possession of reason. The proper use of reason is the measure of man's perfection and fittingness to attain his last end. That is, God did not give us reason so as to act irrationally, but rationally. Otherwise, we become operationally a contradiction in terms: *an irrational rational animal.*

This extends to the use of our various natural powers, since reason dictates that they be used rationally. But the various powers are clearly understood by reason as ordered to certain ends. For example, the power of nutrition is aimed at bodily health through proper eating and drinking. Yet, eating too much or eating poison can damage our health, rendering the nutritional end of the act vitiated. *The act of eating then becomes an anti-nutritional nutritional act, which is self-contradictory, and therefore contrary to rational use of the nutritive faculty.*

Such behavior is irrational, and thus, contrary to man's rational nature. Such behavior therefore leads man away from his true end, which renders such acts something we *ought not do*. It is a thrust away from the true nature of man. Since nature dictates the true being of man, deviating freely from our nature in this fashion is a thrust away from the fulfillment of our being: it is a thrust toward non-being. Thus, it results in a self-destructive act that leads us away from our true end or good.

We overcome the alleged is/ought dichotomy by seeing that immoral acts are committed under the penalty of self-destruction, which is something we should not chose, something we ought not do.

Just as a lame horse suffers physical evil because it lacks the fullness of its natural perfections as a horse, so, too, a man who freely rejects his natural inclination toward moral goodness suffers moral evil for which he is personally responsible.

Such self-destructive acts contradict the creative intention of God in giving us existence and the opportunity to reach our blessed last end. It is a metaphorical "slap in the face to God himself." Natural law tells us that we are not simply "self-responsible," as some claim, but rather are "responsible," not merely to ourselves, but to God as the Supreme Lawgiver and Creator of our human rational natures.

Since reason tells us our immoral acts are contrary to our rational nature, and thus, to God's creative intention, we properly feel guilt, shame, and a realization that we have done something we ought not to do—something that violates the God-given gifts of our existence and rationality.

Without formal realization of the obligation imposed by natural law, all men naturally understand the obligation to do good and avoid evil. They also see to some degree the need to do things that are fitting to their nature. This natural internal compulsion manifests the existence of the natural law in an imperfect way.

Full recognition of natural law is had only by those who also realize that this internal compulsion arises because God exists to impose our nature and its natural ends upon us.

In general, human acts can be understood as morally evil if they entail either a misuse of some natural faculty or the violation of the rights of ourselves or of others. A proper understanding of our powers entails understanding their intrinsic finalities. Thus speech is the means by which truth in my mind is conveyed to another. Lying contravenes that purpose, making communication anti-communicative, and thus, irrational, which, in turn, violates the rationality of our nature.

The violation of others' rights is also seen as irrational, since rights flow from obligations which arise from human nature. For instance, since man must live to fully express God's intention in creating him, to kill him is to violate his right to fulfill his obligation to live so as to reach his last end in God. Equally, suicide violates the right and obligation to maintain our own God-given lives. Without detailing the morality of every act, this is the sort of reasoning which the study of natural law entails.

Clearly, full exposition of all aspects of natural law and its application would require an entire course in ethics, which is impossible in this short paper. Hopefully, some of the above explanations will serve to render clearer the coherency of natural law ethics and its application to current ethical controversies.

Chapter 9

Abortion Ethics: Natural Law vs. Naturalism[1]

This article will examine (1) natural law's and (2) naturalism's opposing views on abortion. Their diverse philosophies determine radically divergent abortion ethics, which will be examined solely through natural reasoning.

Pertinent Thomistic Doctrines

Since embryology teaches that specifically human life begins at conception,[2] modern natural law ethics—following the principles of St. Thomas Aquinas (1225-1274)—prohibits direct abortion at any stage, since it is the taking of innocent human life. This position is consistent with the philosophical doctrine of hylemorphism,[3] which teaches that all physical substances are composed of matter and form. Since form determines what kind of thing a substance is, the human substantial form, or soul, determines the presence of human life. Soul determines that something is a single, unified living organism of a given species.

---

[1] This article first appeared online on the *Strange Notions* website. https://strangenotions.com/abortion-ethics-natural-law-vs-naturalism/

[2] Available online at https://lozierinstitute.org/a-scientific-view-of-when-life-begins/

[3] Available online at https://selfeducatedamerican.com/2012/10/17/the-theory-of-hylemorphism-jonathan-dolhenty/

Embryology makes clear that a specifically human organism, distinct from the body of its mother, begins life at the moment of fertilization. "Zygote" is simply a technical name for the first single-celled human organism. That selfsame human organism lives throughout all the later stages of fetal development, birth, infancy, pubescence, adulthood, and senescence—until death.

Because the same individual human substance lives from conception to death, no change makes it suddenly become a person based on acquisition of certain properties, say, cognitive abilities. *If it is a person at any later stage of life, it is a person at conception and has the same personal rights throughout life.*

Rationality belongs to the human essence throughout life, even though sentient and rational faculties become active as organic functions develop. We do not become human at a certain point of development, only to later lose our human rights because of irreversible dementia. The distinction between potency and act is crucial. The human organism is always a person with rights in act, even though various human faculties may go from potency to act and even back to potency later.

Specifically human faculties (operative potencies) are *not yet in act with respect to their operations*–but are fully in act with respect to existence, even from the time of conception.

These faculties, which are immaterial properties of the human soul, must not be confused with mere brain organization, which, when sufficiently developed, is used for the faculties' operations. *While a certain brain organization is needed for understanding to occur, it is the intellectual faculty that actually allows the person to think.* Mere physical brain activity cannot even perceive sensible images as a whole, and certainly cannot form universal concepts.[4]

This is why faculties must be present from conception, since they have no way of later developing in the immaterial substantial form—and brain function alone can neither think nor sense. The faculties of the soul are needed, so they must begin existence when the soul does.

There has to be a distinction between the substance itself and its operative potencies (powers, faculties). Otherwise, the act of the substance (existence) would be identical to the acts of the faculties (sensing, thinking)—meaning that when not sensing or thinking a man would stop existing!

The faculties exist continuously, while they go into act and sometimes cease acting, and then, begin acting again.

For example, I can be not thinking or seeing, and then begin thinking or seeing, and then cease thinking or seeing again—all the while continuing to (1) exist as a substance and (2) possess the powers to think and to see.

Since this continuously living human organism belongs to the species of rational animal, it is what Boethius defined as a "person," since "a person is a supposit (substance) of a rational nature. The only other such *created* persons are angels.

---

[4] Available online at https://strangenotions.com/how-we-know-the-human-soul-is-immortal/

Natural Law and Abortion

Natural law ethics defines murder as the gravely immoral evil of directly taking innocent human life. Decent men recognize this as a basic moral precept. Without space to defend this precept fully, the argument begins that rights flow from obligations. God gives us life and obliges us to live it well so as to attain the Supreme Good, God himself. That obligation gives us the corresponding right to live. Because others must respect our right to live, it is immoral to violate that right by taking an innocent human life.

Abortion is such an immoral act—evil by its very nature (intrinsically evil), since it directly attacks the most fundamental human right. Nor can such an act be justified by any utilitarian purpose, however good, since *natural law forbids using an intrinsically evil means to attain a good end.*

Up to half of embryos die before implantation.[5] Could their genetic material be so defective that they do not constitute human life? Could one infer that very early stage abortion might be licit? But, those which do survive are human lives and ought not be killed. Those that do not survive either were human or not. If not, then, since they die anyway, there is no reason to kill them. But, if they were human lives, then killing them is clearly immoral. So, the objection is pointless.

But, did not St. Thomas accept the outdated successive animation theory that the human soul is not present at conception, but

---

[5] Available online at https://www.ncbi.nlm.nih.gov/pmc/articles/PMC5443340/

rather appears in the third month, after a vegetative and then sentient soul was present in the first and second months of gestation? Yes, he provisionally accepted Aristotle's reasoning about this, because the ancients did not see how the early stages' matter looked fit for the human form.

Modern Thomists know that the material organization of the human organism, even as a single-celled zygote, is uniquely specific to the human species, as evinced by the uniquely human DNA present in every living human cell, even in its initial stages. Hence, they now correctly insist that the human form must be present, even in the zygote.

Naturalism's Abortion Stance

Accepting neither God nor a spiritual and immortal human soul, naturalism approaches abortion's ethics very differently. Ethical norms themselves are based, not on some transcendental metaphysics, but simply common human approbation of what is right or wrong, possibly augmented by some claim of evolutionary advantage to those who practice such norms. Absent natural law foundations, many different theories are advanced.

For example, the principle not to take innocent human life is not viewed as protective of all human organisms. Rather, criteria as to who merits the "right-to-life-conferring" designation of a "person" are considered and applied only as "warranted."

This means that the human zygote is not considered a person because it lacks certain "personhood" properties, including sentience, self-consciousness, rationality, creativity, socialization, and so forth. Depending upon criteria selected, different stages of fetal development may or may not be granted full human status, with birth being an important event for both ethical and legal purposes.

The gaining of personhood and its corresponding right to life, then, is seen as a gradual process. Properties, such as various levels of cognitive awareness, birth giving "embodiment in the world," and even societal acceptance, become benchmarks by which personhood is more or less arbitrarily conferred upon the developing human organisms.

Using this reasoning, various seeming paradoxes can be explained. For example, *a baby born several weeks prematurely is presently considered a legal person with a right to life, whereas a full term fetus can be aborted at the moment prior to natural birth, since he is still not born—even though neurologically more fully developed than the prematurely born baby.*

One explanation is that birth itself confers "embodiment into the world,"[6] and thus establishes a claim for the premature infant that is lacking to the as yet unborn full term fetus.

For naturalists, the defining characteristic of the human person becomes the development of some brain function that enables activities which fulfill "person conferring" criteria, such as certain cognitive abilities.

---

[6] Available online at https://www.amazon.com/Arguments-about-Abortion-Personhood-Morality/dp/0198806604/

Hylemorphism vs. Atomism

Central is the question of whether the human zygote is correctly described as (1) a hylemorphic unity, that is, living things are made substantially one being by having an immaterial form that unifies the matter and places it into the human species, or (2) as simply a product of biological evolution, that is, a group of synergistically interacting organic molecules with no metaphysical uniting principle.

Because naturalism ascribes no substantial form having special human powers to the human zygote, there is no basis for saying that the zygote is a "rational animal" that would fulfill Boethius's definition of the person, or, for that matter, any other definition of person that naturalists would accept.

Naturalism rejects God's existence, typically by espousing atomism—the claim that all reality ultimately reduces to the smallest physical units: atoms or subatomic particles. Thus, *the focus is now on atomism.*

Does hylemorphism or atomism correctly describe the zygote? Several arguments show that hylemorphism prevails.

First, without hylemorphism, not only is the *substantial unity* of the zygote denied, but so is that of all later stages of human development. Substantial unity means that a thing is undivided in itself and distinct from other beings.

This is crucial because *we do not even exist without such unity.* My video, "Atheistic Materialism,"[7] shows that Richard Dawkins does not exist—based on his own atomistic premises.

If one takes atomism seriously, the only things that really exist are whatever basic atomic or subatomic units of matter one selects as ultimate. For sake of argument, let us consider that the building blocks of organic chemistry and of organisms are atoms. And atoms combine to make larger entities, be they molecules or entire organisms.

The logic is as simple as this: When two atoms combine chemically, say, sodium and chlorine, do they become one thing (salt?), or are they really still two things (two distinct atoms), functionally associated? Atomism logically is forced to the latter position, since all that really exists is atoms, even though they may enter into temporary chemical bonds with other atoms.

The same logic must be followed all the way up through the zygote, the newborn, and the adult human being—including Dr. Dawkins. Atomism is ontologically committed to the sole realities being atoms, despite highly-complex chemical bonding taking place in functional unities (organisms) obeying DNA dictates.

This means that zygotes are not persons, not merely because they lack certain cognitive abilities, but because they lack substantial unity.

In a word, *the philosophy of atomism or naturalism may exist, but atomists and naturalists do not.*

---

[7] Available online at https://youtu.be/rVCnzq2yTCg

Absent some real unifying principle, such as form, atomism's self-defeating truth is that nothing really exists above the atomic level. This is why there is no stable principle of existence on which naturalism can depend to establish a principle of "personhood." Atomism's most embarrassing "public secret" is that, not only is the zygote not a person (on its false premises), but *there is no unified supposit, substance, thing… on which to ground the notion of personhood at all.*

Even the concept of "neural networks or patterns" suffers the same problem. A certain neural pattern may exhibit cognitive activity, much like a computer with AI, but there is no "there" there to be the person having "personhood." Just atoms exchanging outer orbit electrons. That's all.

Atomists have abolished substantial forms. But they have also abolished themselves in the process!

The entire debate over when personhood is present, based on various cognitive or other criteria, misses the point–since the same atomism that denies full human rights to the zygote also denies the ontological basis for the substantial existence of any person at any stage of human life.

The only way to avoid this intellectual suicide is to grant that there is some principle of existential unity above the atomic level in living organisms, including the zygote. Dare we call it a "substantial form?"

What atomism lacks is a stable principle of unification for living organisms. That is why naturalists struggle to designate the point at which the "person" finally appears in fetal development.

"Personhood" never can appear in naturalism, since all really exists is the atoms—not the atomists who believe this fantasy.

At best, the concept of personhood that atomism can support is that of a certain neural pattern that has "achieved" epiphenomenal consciousness. This functionally, but not substantially, unified neural pattern would then become the "person," using the entire organism, brain and all, like a parasite alien invading a victim host.

Even then, the neural pattern's consciousness could not be just material atoms, since even sense experience of the wholeness of images must be immaterial, as shown in an earlier *Strange Notions* piece.[8]

Conversely, if zygotes really exist as substantial unities, hylemorphism reappears—together with continuously existing substances and powers, whose secondary cognitive activities come and go.

Naturalism lacks a continuous principle of substantial unity for the living organism. What hylemorphism has is an essential nature that is present from conception to death—a nature manifesting fully human cognitive abilities sometime after conception.

Previously, I have shown non-material activities of sentient and intellectual substances comport with hylemorphism,[9] but not atomism's materialism—specifically by pointing to (1) sentient cognitive abilities shared by lower animals to apprehend sense data as a whole (which no material device can do), and (2) spiritual intel-

---

[8] Available online at https://strangenotions.com/how-we-know-the-human-soul-is-immortal/

[9] Ibid.

lectual abilities found in man by which he forms universal concepts. Atomism can explain none of these abilities.[10]

Ethical and Legal Principles

Two principles pertain to abortion:

1. It is never licit to directly kill an innocent human being.
2. It is never licit to perform an action unless one is morally certain that it will not directly kill an innocent human being.

These principles have substantial acceptance, both in ethics and in law.

Regarding the first principle, it is the primary obligation of society or the state to protect the innocent human life of every human person within its jurisdiction. With respect to secondary rights, such as bodily autonomy and consent, these rights are limited insofar as they violate the more primary right to life of another human being. Autonomy and consent over my body ends when it allows me to destroy your body and life. That the right to life is more primary than any other right follows from the fact that unless one is alive, no other rights can exist.

The second principle reflects a common sense provision of law. People go to jail for shooting humans they fail to make certain were deer.[11]

---

[10] Ibid.

For hylemorphism, human life begins at conception and is a continuous substantial entity until death. All scientific evidence supports that the same organism present at conception is the one present in adulthood. Therefore, the same substantial form must be present from conception to death. As shown above, this includes the soul's immaterial faculties. Since the form of the adult is clearly human and a person, the soul and human person must have been present from conception on.

Therefore, abortion is always morally wrong and to be condemned. And since it is wrong by its very nature (intrinsically wrong), no set of circumstances or good intentions or out-comes can possibly justify it—for no good end can ever justify an intrinsically evil means to that end.

But what of the view of atomism?

Since neural networks capable of "human" acts, such as creativity, do not develop until a certain point in gestation, or even at birth in the case of "embodiment in the world," atomists claim specifically human faculties are non-existent until such neural development. Thus, they can claim moral certitude (beyond a reasonable doubt) that a "person" does not yet exist, and thus, abortion violates no human rights.

But, as demonstrated above, in atomism the human organism has no substantial unity at any stage of its existence. All that exists are individual atoms. *There is never a substantially unified single being to serve as a substrate for "personhood."*

---

[11] Available online at https://www.yourerie.com/news/local-news/hunter-who-admits-to-accidentally-shooting-and-killing-a-woman-gets-1-3-years/

At best, the "person" turns out to be a certain neural complex of atoms, developed in the human brain, having no substantial unity. Even then, no purely physical neural complex can grasp the wholeness of an extended image in a single, simple act.[12] To do this, an immaterial form is needed, which atomism rejects.

With no immaterial "form" to unite the neural network, its existential experiences of unity and selfhood have no rational basis. Conversely, the existential experience of the self bespeaks some kind of substantial unity, which atomism cannot explain.

Legality

Never even having heard of hylemorphism, the common sense metaphysical insights of most people will lead them rightly to suspect that, if we are a single organism as adults, we must have been one as a human zygote—and that the same individual human life developed from conception into the adult we encounter as a person today.

For that reason, most citizens who follow the ethical principles enunciated above will conclude that one cannot, *with moral certainty*, say at which point the human organism became a person. Being uncertain of this point, they may well conclude that it would be as immoral to kill the unborn child at any point. For this reason, we should not be surprised if legislators conclude that the common ethical principles of society operate to protect the unborn.

---

[12] Available online at https://strangenotions.com/how-we-know-the-human-soul-is-immortal/

Some will protest that society has no right to violate the autonomy and consent of women with pro-life legislation. However, since most people recognize that the first obligation of society is to protect the life of its innocent members, pro-life legislation may find increasing support.

Conclusion

Skepticism today ensnares the minds of many–especially those who have been educated to believe that ancient philosophical and religious belief systems are outmoded and false and that natural science alone offers all possible tentative truths that can be known about man and the cosmos. Modern skepticism denies all traditional value systems, revealed religion, and classical philosophical systems, such as Platonism, Aristotelianism, and Thomism.

Today's dominant intellectual mindset of positivism and materialism begets a worldview that automatically excludes the presence of any truly spiritual realities, and especially, the need for the transcendent God of classical theism. Instead of mere attacks on the more detailed aspects of proofs for God, many recent critics focus on the alleged weakness of the most basic metaphysical principles, such as sufficient reason and causality, on which all such proofs seem to be built. Indeed, analytic philosophy denies to classical metaphysicians even the analogical and transcendental metaphysical first principles of being, claiming that such terms as "existence" or "being" can be predicated only univocally, and thus, could not apply to any transcendent Creator.

For those desiring a deeper understanding of the intellectual foundations which are critically needed in order to defend the key apologetic truths essential to belief in traditional religion and, specifically, Catholicism, this book has provided some of the critical in-sights needed to accomplish that end – written in such a way that the educated layman can understand them.

www.ingramcontent.com/pod-product-compliance
Lightning Source LLC
LaVergne TN
LVHW020935090426
835512LV00020B/3364